THE NAFTA: REPORT ON ENVIRONMENTAL ISSUES

NOVEMBER 1993

For sale by the U.S. Government Printing Office
Superintendent of Documents, Mail Stop: SSOP, Washington, DC 20402-9328
ISBN 0-16-042981-1

THE NAFTA: REPORT ON ENVIRONMENTAL ISSUES

TABLE OF CONTENTS

Page

EXECUTIVE SUMMARY ES-1

I. INTRODUCTION 1

II. NAFTA'S ENVIRONMENTAL PROVISIONS

A. NAFTA 5
B. SUPPLEMENTAL AGREEMENT ON THE ENVIRONMENT 14
C. U.S.-MEXICO BORDER ENVIRONMENT COOPERATION AGREEMENT 22

III. MEXICO'S POLLUTION CONTROL REGIME

A. THE GENERAL ECOLOGY LAW 25
B. ENFORCEMENT OF MEXICAN POLLUTION CONTROL LAWS 35

IV. RECENT DEVELOPMENTS IN THE U.S.-MEXICO ENVIRONMENTAL RELATIONSHIP

A. INTEGRATED BORDER ENVIRONMENTAL PLAN 47
B. TRAINING AND EDUCATION 55
C. CONSERVATION 58
D. FINANCING ENVIRONMENTAL AND CONSERVATION PROGRAMS 65

V. OVERVIEW OF POTENTIAL ENVIRONMENTAL EFFECTS OF NAFTA

A. MACROECONOMIC EFFECTS 69
B. SECTORAL EFFECTS 81
C. EFFECTS ON U.S. ENVIRONMENTAL LAWS 92

VI. SPECIFIC ENVIRONMENTAL EFFECTS

INTRODUCTION AND SUMMARY 100
A. PRODUCT STANDARDS, PESTICIDES, AND FOOD SAFETY 106
B. AIR QUALITY 110

THE NAFTA: REPORT ON ENVIRONMENTAL ISSUES

TABLE OF CONTENTS
(Continued)

Page

VI. SPECIFIC ENVIRONMENTAL EFFECTS (Continued)

C. WATER USE AND QUALITY 120
D. CONTROL OF TOXIC CHEMICALS 126
E. HAZARDOUS WASTE 128
F. NON-HAZARDOUS WASTE 132
G. CHEMICAL EMERGENCIES 135
H. WILDLIFE AND ENDANGERED SPECIES 137
I. FISHERIES 142
J. FORESTS, PARKS AND RANGELANDS 147
K. HEALTH IMPLICATIONS 150

LIST OF TABLES

Table 1 Mexican Adherence to Multilateral Environmental and Conservation Agreements 48

Table 2 Proposed Changes in Border Air Emissions, 1994-2004, Under Three Growth Projections 115

Table 3 Effects on Wildlife and Endangered Species 139

LIST OF FIGURES

Figure 1 Export Growth Since 1987 Has Significantly Boosted U.S. Economic Activity 70

Figure 2 U.S. Exports of Manufactured Goods to Mexico: State Share of Jobs Generated by Export Growth, 1987-92 72

Figure 3 U.S. Investment In Mexico, Total, Direct Foreign, and Domestic, 1992 . . . 73

Figure 4 Rising Income Encourages Lower Birth Rates 75

Figure 5 A Comparison of U.S. and Mexican Environmental Enforcement Efforts 78

EXECUTIVE SUMMARY

INTRODUCTION

The North American Free Trade Agreement ("NAFTA") will create the world's largest market, comprising 370 million people and $6.5 trillion of production. It will open the Mexican economy to U.S. exporters, investors, and service providers, expanding U.S. jobs by an estimated 200,000 in the first two years.

At the same time, the NAFTA is expected to add about 1 per cent to the growth rate of the Mexican economy, accelerating increases in personal income and investments in modern plants, equipment, and infrastructure. This in turn will help create a more economically sound and politically stable neighbor on our southern border, one more able to deal with the environmental challenges that face us.

The promotion of trade and investment throughout the continent under the NAFTA has potential effects on the physical environment and on environmental policies and programs. These potential effects were taken into consideration in negotiating the NAFTA. When President Clinton endorsed the NAFTA in October 1992, he recognized that the NAFTA itself contained provisions to protect the environment. He committed, however, to pursue additional measures, including supplemental agreements, to ensure that increased trade with Mexico and Canada would not come at the expense of the environment.

As part of this commitment, the President recently signed the North American Agreement on Environmental Cooperation with the governments of Canada and Mexico. The fundamental objectives of this Agreement, which will come into force when the NAFTA goes into effect, are to promote cooperation to improve environmental conditions throughout North America and to improve national enforcement of each country's laws relating to environmental protection.

In addition, the Clinton Administration recently announced completion of negotiations with Mexico on an agreement to establish two new institutions devoted to environmental improvement in the border area: the Agreement on the Border Environment Cooperation Commission and the North American Development Bank. The Border Environment Cooperation Commission will work with local communities to develop and arrange financing for vitally needed environmental infrastructure projects. The North American Development Bank will use 90 per cent of its capital -- to be contributed equally by the United States and Mexico -- to leverage some $2 billion or more of private funds in the capital markets in order to finance the construction of these border environmental projects through bonds and other instruments.

Even with these steps, the potential impact of the NAFTA on the environment continues to create misgivings or misunderstandings among some about the agreement. Because of the high priority the Clinton Administration places on protecting the environment, and on the importance of informing the public fully on the issues, the Office of the U.S. Trade Representative ("USTR") coordinated this report on the likely significance of the NAFTA and the side agreements on environmental and conservation issues.

The Administration firmly believes, and this Report shows, that the NAFTA and its related environmental agreements establish precedent-setting international mechanisms and national commitments that will make us more effective than ever in protecting the environment of the United States and all of North America.

NAFTA AND THE SIDE AGREEMENTS

Provisions of the NAFTA itself ensure that the United States can maintain and enforce its existing federal and state health, safety, and environmental standards, as well as U.S. international treaty obligations to limit trade in controlled products such as endangered species.

In addition, NAFTA expressly endorses the principle of sustainable development and includes environmentally sensitive provisions on dispute settlement and investment. As a result, NAFTA goes further than any previous trade agreement in addressing environmental concerns associated with the elimination of trade barriers.

The North American Agreement on Environmental Cooperation (Environmental Agreement) supplements the environmental provisions and objectives of the NAFTA to maintain and enhance environmental protection. The Environmental Agreement establishes a framework for United States, Mexican and Canadian cooperation on a long agenda of common environmental matters. The Agreement specifically commits the parties to effective enforcement of their environmental laws. It also provides for monitoring the environmental effects of the NAFTA. The creation of this unprecedented institution and the parties' ability to hold each other accountable for the agreement's environmental commitments hinges, however, upon the entry into force of the NAFTA.

To develop the environmental infrastructure of the border region, the United States and Mexico have reached an additional side agreement to establish two new institutions: a Border Environment Cooperation Commission ("BECC") to help local communities plan and coordinate environmental infrastructure projects; and a North American Development Bank ("NADBank") to help generate financial resources for their construction.

MEXICO'S POLLUTION CONTROL REGIME

In order to address "pollution haven" and transboundary pollution concerns, in 1993 the U.S. EPA carried out a comparison of U.S. and Mexican environmental laws, regulations and standards. This comparison covered water, air, hazardous waste, pesticides, and industrial chemicals, among other topics. As a result of this study, EPA has concluded that, overall, the United States and Mexican regulatory regimes are designed to achieve comparable levels of environmental protection.

In 1992, Mexico restructured its federal environmental program and made it a major component of its Secretariat for Social Development (SEDESOL). The creation at that time of a new semi-independent office for environmental enforcement, the Federal Attorney General for Environmental Protection ("PFPA"), marked a significant change in the development of its vigorous program of inspections, leading to increasingly tough enforcement follow-up when violations are discovered.

Recognizing that Mexico's General Ecology Law was only enacted in 1988, and that additional technical sophistication in the inspection program will help bridge the gap between establishment of a strong enforcement presence and widespread development of compliance--oriented environmental management practices by industry, Mexico has made impressive strides in implementing its enforcement program in just five years.

RECENT DEVELOPMENTS IN THE U.S-MEXICO ENVIRONMENTAL RELATIONSHIP

The United States and Mexico have a long history of cooperation on environmental issues. These joint efforts were given added impetus by the NAFTA negotiations and the negotiation of the related environmental agreements. Implementation of the NAFTA can be expected to further strengthen the commitment of both the U.S. and Mexican governments to these programs.

The Integrated Environmental Plan for the Mexico-U.S. Border Area ("Border Plan"), which was issued in February 1992, is intended to provide for the long-term protection of human health and the environment within the U.S.-Mexico border area. The Plan contemplates a multi-phase process to achieve this goal. The Border Plan has the following four principal objectives:

(1) to strengthen enforcement of existing environmental protection laws;

(2) to reduce pollution and improve the quality of the border area through new initiatives;

(3) to increase cooperative planning, training and education; and,

(4) to improve understanding of the border area environment.

Since 1992, EPA and SEDESOL have implemented a wide range of cooperative projects to fulfill these four objectives.

At their meeting in Ensenada, Mexico, in October 1993, EPA Administrator Browner and SEDESOL Secretary Colosio announced plans to begin work in early 1994 on a more comprehensive Border Action Program to continue and expand the work under the Border Plan. They also committed to begin work promptly on joint efforts to study visibility impairment in the border area that may stem from power plants (such as Carbon I and II) and other emission sources in both countries, and to develop appropriate control measures under the La Paz agreement on border environmental cooperation.

During the last year and a half, EPA, the U.S. Fish and Wildlife Service and the National Oceanic and Atmospheric Administration have worked with SEDESOL and other Mexican agencies on a number of cooperative training and education activities. These activities have focused on environmental enforcement, pollution monitoring and control, and management of nature reserves and marine resources. While many of these programs have been designed for Mexican environmental officials, other programs have been targeted at maquiladoras, local community officials, and the public.

By virtue of their long common border, migration patterns, and distribution of many species, both countries have a common interest in a great number of wildlife and natural resource issues. The entire border region, and in particular the Rio Grande Valley, has a great deal of unique biological diversity which the two sides are taking steps to protect. The United States and Mexico have a long history of cooperating on wildlife protection and the conservation of natural resources. The NAFTA, the Environmental Agreement, and the plans for the new Border Action Program build the base for closer cooperation in the future.

Environmental protection programs cost money, which often means public funds that are subject to heavy demands for many needs, particularly in developing countries. Financial mechanisms targeted directly to environmental infrastructure projects or major conservation programs can provide needed financial support for efforts to protect and improve the environment. The recently negotiated U.S.-Mexico Border Environment Cooperation Agreement to establish the Border Environment Cooperation Commission and the North American Development Bank, represents one such environmental financing mechanism directly related to the NAFTA. Other national and international sources of environmental financing include debt-for-nature swaps and multilateral development bank loans to Mexico.

MACROECONOMIC EFFECTS

Implementation of the NAFTA is expected to promote additional development of both the U.S. and Mexican economies. In the short run, this economic development will take place through the increased specialization and economies of scale resulting from the freer flow of trade in goods and services. In the long run, the expansion of research and development, training, and capital equipment encouraged by the NAFTA will further enhance economic activity. For Mexico, improved access to U.S. technology in the energy, communications, electronics, and other critical industries should provide an additional boost to growth.

The economic changes that will come with the NAFTA have the potential to place additional stress on the environment, particularly for the development of transport and other trade-related facilities at border crossings, but they will also expand the public and private resources for pollution control and other environmental efforts and disperse industrial development away from the already stressed border area of Mexico. Thus, the NAFTA and the side agreements have clear environmental benefits on a national scale and over the long term, but may result in some adverse local impacts in the short term. These specific effects are discussed in Section VI of the Report.

Defeat of the NAFTA would not perpetuate the status quo. Rather, it would be almost certain to reduce U.S. exports to Mexico. Mexico could suffer from a loss of confidence in its economic growth prospects, reduced domestic and foreign investment, and slower growth. While Mexico would surely exercise policy options in an attempt to preserve its prospects for economic development, these would be limited, at least in the short run. The near-term effects of NAFTA's defeat in Mexico would be less growth and more poverty. In the United States, this would result in reduced exports to Mexico, fewer high-paying U.S. jobs supported by such exports, and, ultimately, higher levels of immigration from Mexico thatn would otherwise be the case.

Available evidence and empirical research suggest that environmental considerations are generally not important determinants of investment decisions in North America. Furthermore, under the NAFTA and the Environmental Agreement, each country undertakes to avoid environmental policies that would divert investment from one country to another. Finally, NAFTA will remove the current artificial incentives which have intensified investment along the border through the maquiladora program. Without NAFTA, it is more likely that intense border investment will continue, with the attendant adverse environmental consequences for the border region.

SECTORAL EFFECTS

Energy:

Concern for a more efficient, less vulnerable, and environmentally sustainable energy future was the foundation of the negotiations of the energy provisions of NAFTA. By removing barriers and disincentives to the use of natural gas in Mexico, NAFTA opens up prospects for cleaner power generation. This, together with greater scope for private investment in Mexico's power generation sector, should help mitigate environmental effects of increased demands for electricity created by economic growth.

By aligning Mexican oil and gas prices with market forces, NAFTA will open greater opportunities for access to renewable energy technologies. NAFTA also allows continuation of incentives for exploration and development of non-renewable energy sources, but does not change disciplines on subsidies spelled out elsewhere in the Agreement.

Absent NAFTA, the benefits of greater use of natural gas and of cleaner technologies could be lost. If the NAFTA is not approved, U.S. opportunities to market goods and services in the Mexican economy would be substantially reduced and the environmental benefits from increased sales of U.S. energy goods and services, including natural gas and renewable energy technologies, would be significantly curtailed.

Agriculture:

NAFTA is expected to result in increases of U.S. exports of grains, oilseeds, and meats, while fruits and vegetables are likely to account for the bulk of Mexico's increased exports. Effects on environmental quality depend in part on the resultant changes in the output and input mix, and also on land use patterns. Chemical use changes will differ regionally, but an overall net result is difficult to discern. Mexico's corn market liberalization, and reduction in water and chemical input subsidies, could result in reduced cultivation of marginal lands, reducing potential for erosion, deforestation and loss of biodiversity.

Transportation:

By promoting economic growth and increased trade, NAFTA will heighten transport-related environmental concerns such as congestion, noise, and emissions. However, NAFTA also contains offsetting provisions that address these same concerns, such as opening up cross-border trucking to avoid delays, and eliminating the need for inefficient "empty" return trips and risky transfers of hazardous cargoes at the border. All trucks operating in the U.S. market will be required to meet U.S. safety standards. NAFTA also provides a vehicle for upward harmonization of safety and emission standards through the work of the Automotive Standards Council. Recently introduced efficiencies in rail links should facilitate intermodal carriage and alleviate some of the added pressures on the trucking sector.

The NAFTA also provides specific customs administration procedures to streamline import and export procedures and to ease congestion at customs border points. This will alleviate air pollution from vehicles in U.S.-Mexican border cities. NAFTA's Customs Administration provisions establish a Working Group, which will meet regularly to address ways to facilitate trade flows, including harmonization of automation requirements and documentation, and proposed administrative and operational changes.

U.S. Environmental Technology and Services:

With the passage of NAFTA, Mexican demand for environmental technologies will increase significantly, presenting substantial commercial opportunities for the U.S. environmental technologies industry and creating high-wage jobs for American workers. Implementation of NAFTA and the associated Environmental Agreement is expected to increase considerably environmental spending in Mexico -- from $2 billion in 1993 to well over $3 billion by 1997.

EFFECTS ON U.S. ENVIRONMENTAL LAWS

Much of the concern about NAFTA and the environment has focused on whether NAFTA will undermine the ability of the U.S. government and the states to establish and enforce their environmental, health, and saftey laws. In fact, far from weakening environmental, health, and safety measures, the NAFTA and the Environmental Agreement affirmatively encourage the three countries to improve standards and strengthen enforcement.

SPECIFIC ENVIRONMENTAL EFFECTS

Product Standards, Pesticides and Food Safety:

NAFTA's provisions ensure that the ability of the United States to establish and enforce its food safety and pesticide standards will be maintained and that the integrity of U.S. regulatory processes will be fully respected. Through the NAFTA and the Environmental Agreement, enhancement of product standards and enforcement activities will be promoted in all three countries. NAFTA also provides strong incentives and an excellent opportunity to share expertise and experience to secure real public health and environmental gains.

Air Quality:

With or without the NAFTA, increasing industrialization and population growth is likely to lead to increased air emissions in the U.S.-Mexico border area. Implementation of the NAFTA will provide added impetus to cooperative programs already underway between the U.S. and Mexican governments to improve air quality in the border region. Through such cooperation, within eight to ten years of implementation of the NAFTA it is even possible that border area air emissions in Mexico could be reduced below current levels if the most favorable projections of cooperative efforts and dispersal of growth in Mexico are realized.

Without NAFTA, the continuation of the maquiladora program is likely to result in more rapid growth in air emissions in the border region, since incentives will be maintained for companies to build additional facilities near the U.S.-Mexico border. Such facilities are likely to be associated with further population increases, which will, in turn, contribute additional sources of air pollution in the form of motor vehicle emissions and pollution from residential heating fuels.

Water Quality and Supply:

Continued growth and development in the U.S.-Mexico border region--which could occur with or without NAFTA--will place increasing demands on already scarce water resources. There are three main areas of concern: (1) addressing limitations on surface water and groundwater supplies; (2) improving water quality by constructing adequate wastewater treatment facilities; and (3) preserving ecosystems, wildlife habitats and coastal areas.

Implementation of the NAFTA will provide added impetus to cooperative projects already under way between the United States and Mexico, pursuant to the Integrated Border Environmental Plan, to promote water quality and preserve the border environment. Moreover, the Border Environment Corporation Agreement will provide additional financing for infrastructure projects to treat wastewater and provide clean drinking water supplies.

Without NAFTA, the United States and Mexico may have difficulty sustaining their current level of cooperation in projects designed to maintain water quality. In addition, it may be difficult to obtain sufficient financing to design and construct needed wastewater treatment facilities in the border area. Moreover, in the absence of NAFTA, incentives for companies to locate facilities in the border region will continue, resulting in additional strains on already-scarce water resources.

Control of Toxic Chemicals:

The NAFTA maintains the ability of the United States to control imports of toxic chemicals from Canada and Mexico, pursuant to the Toxic Substances Control Act. In addition, the NAFTA includes provisions which encourage toxic chemical data and studies generated in Mexico to meet the same standards and Good Laboratory Practices as data generated in the United States.

Hazardous Waste:

With or without NAFTA, economic growth in the United States and Mexico is likely to generate increasing amounts of hazardous waste. With the NAFTA, however, resources will be made available to manage such waste properly and encourage enforcement of hazardous waste laws.

Article 104 of the NAFTA assures that U.S. agreements with Canada and Mexico to manage cross-border shipments of hazardous waste will prevail over NAFTA's obligations if there is an inconsistency. Article 104 makes the same provision for the Basel Convention, which governs international shipments of hazardous waste.

In addition, Chapter 12 of the NAFTA ensures that U.S. environmental engineering and waste management firms and professionals will be able to offer their services in Mexico to construct safe disposal facilities for hazardous waste.

In the absence of NAFTA, the amount of hazardous waste from Mexico requiring management in the United States could increase, as a result of several factors. First, the continuation of the maquiladora program is likely to result in increasing industrial activity in the border area with associated growth in generation of hazardous waste. In contrast, implementation of the NAFTA could significantly reduce the incentives to locate in this region. Second, without NAFTA and its associated Border Environmental Cooperation Agreement, Mexico may have fewer funds to invest in waste management infrastructure. Finally, without the NAFTA, the Environmental Agreement's mechanisms to foster enforcement efforts will not go into effect.

Non-hazardous Waste:

With or without the NAFTA, economic growth in the United States and Mexico is likely to generate increasing levels of non-hazardous wastes. The lack of infrastructure and proper waste management practices along the border is already causing serious environmental and health concerns. New waste generation will only add to these problems.

The Border Financing Agreement will give a preference to infrastructure projects addressing solid waste disposal needs. Moreover, implementation of the NAFTA may reduce the growth of waste along the U.S.-Mexico border by decreasing incentives for industries and populations to locate along the border.

In the absence of NAFTA, the continuation of the maquiladora program is likely to result in increasing population and industrial growth rates along the U.S.-Mexico border. Further, without NAFTA it may be very difficult to secure adequate funding to properly manage the waste generated in this region.

Chemical Emergencies:

Increased industrial growth near the U.S.-Mexico border -- which could occur with or without NAFTA -- may increase the likelihood of a chemical emergency affecting the environment of both countries.

The United States and Mexico are engaged in cooperative efforts to improve emergency preparedness coordination between the two countries. The Supplemental Agreement on the Environment specifically includes "environmental emergency preparedness and response activities" as an item in the work program of the Council for Environmental Cooperation.

Wildlife and Endangered Species:

Current development and activities in the U.S.-Mexico border area are having an adverse impact on wildlife and endangered species in the border area.

Although NAFTA could contribute to short-term slight to moderate increases in these adverse effects as a result of increased trade and development between the two countries, in the long term, increased opportunities for cooperation between the United States and Mexico as the result of the NAFTA will help to address the stresses of development. Furthermore, maquiladora development will tend to be dispersed away from the border area to other parts of Mexico, thus reducing its impact on the border area. Finally, the Environmental Agreement provides a mechanism to address many of these problems, especially those related to enforcement. In particular, new environmental funding and increased personnel could result in improved environmental conditions and reduced environmental effects in the border regions of both countries.

If NAFTA is not implemented, incentives will continue under the maquiladoras to locate facilities in the border areas, thus exacerbating environmental pressures on the border, such as loss of habitat, adverse impacts to endangered and threatened species, and reductions in groundwater levels. Moreover, many of the increased opportunities for cooperation, training and enforcement of wildlife protection laws discussed below would be lost.

Fisheries:

Although some fisheries-related environmental problems (in particular, the tuna-dolphin and shrimp-turtle problems) have been a source of friction between the United States and Mexico, substantial progress has been made by Mexico to address these problems. Indeed, Mexico has reduced its dolphin mortality rates dramatically.

NAFTA provides opportunities for increased cooperation in fishery management and conservation, and safeguards enforcement of laws relating to use of these resources, for example through the CEC established under the Supplemental Agreement. This could result in improved management of each species and conservation of depleted stocks.

Without NAFTA, current bilateral consultations will continue, but without the improved atmosphere and high-level attention that the Environmental Agreement provides. NAFTA has provided the incentive for many significant improvements in Mexico's conservation policies regarding living marine resources. Without it, the United States will not have a commitment from Mexico to give effect to these policies.

Forests, Parks and Rangelands:

While deforestation in Mexico will likely continue, it is anticipated that NAFTA could slow the rate of deforestation somewhat, due to a general increase in wage rates and an increase in alternative sources of productive employment.

In the short term, resource threats to the National Park System may be created by an increase in cross border transportation, increased population settlement with some additional industrial site development at or near the border, consequent decreased water quality, and increased destruction of cultural resources. However, increased opportunities for joint cooperation to address these problems may alleviate some of their impacts. NAFTA may increase demand for services from public lands, but otherwise its impact on them will be negligible.

If NAFTA is not implemented it is unclear whether alternate mechanisms to reduce deforestation exist. Pressure in the border areas would continue, particularly those driven by the concentration of maquiladoras. Without NAFTA, public use and demands on public lands will continue, but likely at lower levels than with the implementation of NAFTA. However, without NAFTA, the level of adverse impacts to park resources would be slightly greater than long-term conditions with NAFTA due to less joint coordination and cooperation between U.S. and Mexican environmental agencies to reduce environmental problems.

Health Implications:

Assuming that the NAFTA is successful in increasing economic development in Mexico generally, and along the border specifically, increased governmental support to the community and health infrastructure is expected, accompanied by an improved overall health status.

In the absence of NAFTA, if Mexico does not place greater emphasis on environmental enforcement along the border, the environmental health issues could increase significantly, particularly for victims of surface water and groundwater pollution. Health problems ranging from bacterial infections to toxic effects from hazardous waste could also increase. Health and environmental resources, currently inadequate on both sides of the border, could be expected to remain so.

I. INTRODUCTION

The North American Free Trade Agreement ("NAFTA") will create the world's largest market, comprising 370 million people and $6.5 trillion of production. It will open the Mexican economy to U.S. exporters, investors, and service providers, expanding U.S. export-supported jobs by an estimated 200,000 in the first two years. At the same time, it may add up to 1 percent to the growth rate of the Mexican economy, accelerating increases in personal income and investments in modern plants, equipment, and infrastructure.

The promotion of trade and investment throughout the continent under the NAFTA has potential effects on the physical environment and on state and national environmental policies and programs. When President Clinton endorsed the NAFTA in October 1992, he recognized that the NAFTA itself contained provisions to protect the environment; however, he committed to pursue supplemental agreements to ensure that increased trade with Mexico and Canada would not come at the expense of the environment. As part of this commitment, the President recently signed the North American Agreement on Environmental Cooperation with the governments of Canada and Mexico. The fundamental objectives of this agreement, which will come into force when the NAFTA goes into effect, are to promote cooperation to improve environmental conditions throughout North America and to improve national enforcement of each country's laws relating to environmental protection.

In addition, the Clinton Administration recently announced completion of negotiations with Mexico on a Border Environment Cooperation Agreement. This agreement establishes two new institutions devoted to environmental improvement in the border area. The Border Environment Cooperation Commission will work with local communities to coordinate, develop, and arrange financing for vitally needed environmental infrastructure projects. The North American Development Bank will use 90 percent of its capital -- to be contributed equally by the United States and Mexico -- to leverage some $2 billion or more of private funds in the capital markets in order to finance the construction of these border environmental projects through bonds and other instruments.

Even with these steps, the potential impact of the NAFTA on the environment continues to create misgivings towards the agreement, which are sometimes fed by misunderstandings or misinformation. Because of the high priority the Clinton Administration places on protecting the environment, and on its desire to inform the public fully on the issues, the Office of the U.S. Trade Representative ("USTR") has coordinated this Report on the likely significance of the NAFTA and the side agreements on environmental and conservation issues.

This Administration firmly believes, and this Report shows, that the NAFTA and its related environmental agreements will establish precedent-setting international mechanisms and national commitments that will make us more effective than ever in protecting the environment of the United States and all of North America.

The 1992 Environmental Review

Even before the NAFTA negotiations began, as Congress considered granting the President negotiating authority in 1991, concerns were raised about the potential environmental effects of such an agreement. Some feared that a free trade agreement with Mexico would aggravate environmental problems on the border. Some sought assurances that the trade agreement would not weaken U.S. health, safety and environmental standards. Some expressed concern that Mexico's environmental laws were inadequate and their enforcement so lax that Mexico would become a "pollution haven" for U.S. companies trying to avoid environmental regulation. Some worried that U.S. laws to protect wildlife could be undermined.

In May 1991, in response to these concerns, President Bush committed his Administration to undertake a review of U.S.-Mexico environmental issues, including an analysis of the possible environmental effects of a free trade agreement. The review was undertaken at the outset of the negotiations, so that the information obtained could be taken into account by U.S. officials during the NAFTA negotiations. A first draft of the review was made available to the Congress, the general public and U.S. NAFTA negotiators on October 17, 1991. USTR received public comments on the draft review through January 30, 1992. The review was revised to include a general discussion of the comments, and some changes were made to the review in response to specific comments. The final 231-page report, Review of U.S.-Mexico Environmental Issues, was released in February 1992.

The 1992 Environmental Review concluded that NAFTA would have significant potential environmental benefits and limited adverse effects for the environment in the United States and Mexico. This conclusion was based on four major factors:

(1) The Review found that the experience of other countries showed that increased economic growth generated by NAFTA would generate greater domestic demand for improved environmental quality and provide Mexico with additional motivation and resources to invest in environmental protection.

(2) The Review also determined that NAFTA would ease environmental pressures on the U.S.-Mexico border, as free trade would encourage economic development to occur beyond the border region.

(3) The Review concluded that Mexico would not become a pollution haven for U.S. firms for several reasons. First, investors could not expect a significant difference over time between U.S. and Mexican environmental requirements. Second, pollution abatement costs represent a small share of production costs for most U.S. industries. Finally, many of the industries with high environmental costs are capital intensive, making it economically impractical for them to move to a new location.

(4) Finally, the Review identified a number of other environmental benefits from NAFTA, such as improving Mexican access to clean fuels and technology, and easing environmental pressures in Mexico City.

An important purpose of the Review was to provide guidance to U.S. negotiators in achieving an environmentally sound agreement. Consequently, the review set forth a series of recommendations that "would help NAFTA have an environmentally positive impact." Most of the recommendations were subsequently incorporated in the NAFTA at the request of the U.S. negotiators.

Scope of This Report

The Clinton Administration is committed to ensuring that Congress and the public have a full and fair understanding of the environmental issues that arise in connection with the NAFTA. The Administration wants Congress and the public to know the extent to which environmental considerations were taken into account in the negotiation of the NAFTA and the side agreements, as well as the results of those negotiations. To this end, the Office of the U.S. Trade Representative has prepared this report to provide a balanced assessment of the likely effects on the salient environmental issues raised by the NAFTA.

In preparing this Report, the Administration has focused on new developments since the 1992 Environmental Review. Most notably, of course, the NAFTA itself was concluded, with a number of provisions which specifically address environmental issues. In addition, the Clinton Administration concluded the North American Agreement on Environmental Cooperation with Canada and Mexico, and has negotiated the U.S.-Mexican Agreement on the Border Environment Cooperation Commission and the North American Development Bank with Mexico. This Report also addresses environmental concerns that have been raised since February 1992. Finally, where new information is available, analyses performed for the 1992 Review have been updated.

Most of the Report focuses on environmental issues between the United States and Mexico. Because a free trade agreement is already in effect between the United States and Canada, implementation of the NAFTA's trade provisions should have few environmental implications. Moreover, in October 1992, the Government of Canada issued the Canadian Environmental Review on environmental implications of the NAFTA for Canada.

This Report was prepared by an interagency task force, coordinated by the Office of the U.S. Trade Representative. The task force included representatives from the Departments of State, Treasury, Commerce, Transportation, Agriculture, Justice, Interior, Health and Human Services, Energy, and Labor; the Environmental Protection Agency; the Council of Economic Advisors; and the Office of Management and Budget.

The Report comprises six sections:

- Section I describes the background and scope of the Report.
- Section II outlines the environmental provisions of the NAFTA, and describes the Environmental Agreement and the Border Environment Cooperation Agreement.
- Section III surveys Mexico's General Ecology Law and its recent environmental enforcement efforts.
- Section IV describes recent developments in the U.S.-Mexico environmental relationship.

- Section V provides an assessment of the likely effects of the NAFTA on such broad issues as economic growth, investment flows, various economic sectors, and U.S. environmental laws.

- Section VI analyzes the potential effect of the NAFTA on specific environmental issues, such as water quality and trade in endangered species.

The Clinton Administration recognizes that there are serious environmental problems in North America, particularly along the U.S.-Mexican border, that must be addressed in the context of expanded trade. Greater economic integration among the United States, Canada and Mexico calls for new commitments to intensify our cooperative efforts to protect and enhance the quality of the environment, and new institutions to make those commitments a reality. The question that Congress now faces, and that this Report attempts to answer, is whether passage of the NAFTA will exacerbate our environmental problems or give us effective mechanisms to ameliorate them.

II. NAFTA'S ENVIRONMENTAL PROVISIONS

This Section describes the contents of the three NAFTA documents relating to the environment:

- the environmental provisions of the NAFTA itself;
- the North American Agreement on Environmental Cooperation ("the Environmental Agreement"); and
- the U.S.-Mexico Agreement on the Border Environment Cooperation Commission and the North American Development Bank ("the Border Environment Cooperation Agreement").

The effects that these agreements will have on the environment are addressed in greater detail throughout other relevant Sections of this Report. (See, e.g., Section V.C, "Effects on U.S. Environmental Laws.")

A. NAFTA

The provisions of the NAFTA itself ensure that the United States can maintain and enforce its existing federal and state health, safety, and environmental standards, as well as U.S. international treaty obligations to limit trade in controlled products, such as endangered species. In addition, NAFTA expressly endorses the principle of sustainable development and includes environmentally sensitive provisions on dispute settlement and investment. As a result, NAFTA goes further than any previous trade agreement in addressing environmental concerns associated with the elimination of trade barriers.

1. Endorsement of the Principle of Sustainable Development

The NAFTA parties have agreed that economic development should take place in an environmentally sound manner. Thus, the NAFTA Preamble states that one of its primary purposes is to:

> "Contribute to the harmonious development and expansion of world trade ... in a manner consistent with environmental protection and conservation; ... promote sustainable development; ... [and] strengthen the development and enforcement of environmental laws and regulations."

2. Health, Safety and Environmental Standards Provisions

The NAFTA includes agreements on sanitary and phytosanitary measures and standards-related measures. Sanitary and phytosanitary measures ("SPS") generally deal with protecting human, animal and plant life and health from risks of plant- and animal-borne pests and diseases, as well as with additives and contaminants in foods and feedstuffs. Examples of such measures include U.S. pesticide residue tolerances and restrictions on food additives. Standards-related measures ("SRMs") deal with voluntary and mandatory product standards and the procedures used to determine whether a particular product meets a standard. Examples of SRMs include toxic chemical testing requirements, and U.S. regulations concerning pharmaceuticals and other health care products.

The NAFTA explicitly recognizes that countries have a legitimate need for product standards and for regulations to protect human, animal and plant life and health. The NAFTA provisions are designed to preserve the ability of governments to act in this area while guarding against the unjustified use of these types of measures to protect domestic industry. In each case, the NAFTA sets up requirements and procedures that will help to distinguish legitimate measures from primarily protectionist measures. The NAFTA also will help facilitate efforts to make these measures compatible among the three NAFTA parties, where appropriate.

The NAFTA text on sanitary and phytosanitary measures (Chapter Seven, Section B) differs from the NAFTA text on standards-related measures (Chapter Nine) in the means used to determine whether a measure is protectionist in nature. As described below, Chapter Nine relies on non-discriminatory treatment, while Section B of Chapter Seven relies on science and risk assessment. A strict requirement for non-discriminatory treatment is not possible for sanitary and phytosanitary measures, since such measures must frequently discriminate against imported goods or goods from a particular country because those goods pose a different risk of a plant or animal pest or disease. Under Section B, discrimination is allowed as long as it is not arbitrary or unjustifiable.

NAFTA has very little bearing on most U.S. environmental laws and regulations. While some environmental standards are SPS measures as defined in Chapter Seven, or standards-related measures as defined in Chapter Nine, the great majority are not covered by the NAFTA. For example, several key U.S. environmental regulatory statutes (including the Clean Air Act, the Clean Water Act, the Resource Conservation and Recovery Act) are generally neither SPS nor standards-related measures and are not within the scope of the NAFTA.

a. Summary of provisions

The NAFTA chapters on Standards-Related Measures and Sanitary and Phytosanitary Measures preserve the ability of the United States to maintain, strengthen, and enforce existing U.S. health, safety, and environmental standards, and encourage our NAFTA trading partners to strengthen their standards. Specifically, the NAFTA:

- Affirms the right of each party to choose the level of protection of human, animal, or plant life or health it considers appropriate;

- Maintains existing U.S. federal and state health, safety, and environmental standards, and preserves the right to ban non-conforming imports;

- Continues to allow the parties, including states and cities, to enact standards that are stricter than international or national standards;

- Commits the NAFTA parties to work jointly to enhance their standards through upward harmonization;

- Continues to allow parties to act to protect human, animal or plant life or health based on available information when there is insufficient information to conduct a risk assessment;

- Ensures the ability of each country (and interested groups within them, including non-governmental organizations) to receive advance notification of proposed regulatory actions in the other two countries, to review and comment upon those actions, and to have such comments taken into account prior to final decision;

- Establishes a Committee on Standards-Related Measures to facilitate compatibility of standards, consult regularly on matters of common concern in this area, and enhance cooperation on developing, applying, and enforcing standards-related measures; and

- Establishes a Committee on Sanitary and Phytosanitary (SPS) Measures to enhance food safety and improve sanitary conditions, promote harmonization and equivalence of SPS measures, and facilitate technical cooperation and consultation on specific SPS issues.

b. Chapter Seven: Sanitary and phytosanitary standards ("SPS")

Following is a description of the key SPS provisions of Chapter Seven. The Statement of Administrative Action includes a more extensive discussion of the purpose and meaning of these provisions.[1]

Article 712 -- Right to take SPS measures. Article 712.1 explicitly recognizes the rights of governments, including state and local governments, to adopt, maintain or apply sanitary and phytosanitary measures, including measures more stringent than an international standard.

Article 724 -- Appropriate level of protection. In taking sanitary and phytosanitary measures to protect against risks, a central question is how much protection the government seeks to provide against a particular risk. Under the NAFTA, the answer to this question is left up to each government. Each government decides on the "appropriate level of protection,"

---

[1] The Statement of Administrative Action is the document prepared by the Administration and transmitted to Congress for Congressional approval along with the NAFTA implementing legislation that, among other things, describes the NAFTA provisions and explains how the implementing bill and proposed administrative actions change or affect existing law. This Statement is required pursuant to Section 1103 of the Omnibus Trade and Competitiveness Act of 1988, and should be viewed as the authoritative Administration interpretation of the NAFTA's provisions.

which is defined in Article 724 as "the level of protection of human, animal or plant life or health in the territory of a Party that the Party considers appropriate."

Governments are free to choose this level of protection; the NAFTA explicitly affirms this right (Article 712.2). In the end, the choice of the appropriate level of protection is a social value judgement. There is no requirement for a scientific basis for the level of protection, because it is not a scientific judgement.

Article 713 -- Use of international standards. Recognizing that the sanitary and phytosanitary measures of the three NAFTA countries are often different, Article 713 provides for the use of relevant international standards, without reducing the level of protection of human, animal or plant life or health, as a basis for each NAFTA government's own sanitary and phytosanitary measures. The objective in using international standards is to facilitate trade by making the NAFTA parties' measures equivalent or, where appropriate, identical. At the same time, as noted above, the NAFTA explicitly affirms the right of each government to have a sanitary or phytosanitary measure more stringent than the relevant international standard, as long as the sanitary or phytosanitary measure is consistent with the other SPS provisions.

The NAFTA provisions on sanitary and phytosanitary measures were specifically negotiated to make clear that there would be no "downward harmonization" of such measures. While governments are required to use international standards as "a basis" (but by no means the only basis) for their sanitary and phytosanitary measures, Article 713.1 explicitly states that this is to be done "<u>without</u> reducing the level of protection of human, animal or plant life or health." (Emphasis added.) Article 713.3 also explicitly provides that <u>nothing</u> in this requirement "shall be construed to prevent a Party from adopting, maintaining or applying, in accordance with the other provisions of this Section [concerning, for example, not arbitrarily discriminating between its goods and like goods of another Party], a sanitary or phytosanitary measure that is more stringent than the relevant international standard, guideline or recommendation."

Article 714.1 -- Equivalence. In order to minimize the differences between the three NAFTA countries' sanitary and phytosanitary measures, Article 714.1 provides procedures to make the parties' measures equivalent. However, the text makes clear that these procedures must not reduce the level of protection of human, animal or plant health, or of the environment. Although an importing party must accept a measure of an exporting party as equivalent if the exporting party's measure achieves the importing party's appropriate level of protection, the importing party has the right to determine that the measure does not achieve its appropriate level of protection, where it has a scientific basis for such a determination.

Article 718 and 719 -- Transparency. Article 718 requires advance public notice and opportunity to comment on proposed sanitary and phytosanitary measures or modifications to sanitary and phytosanitary measures. In the case of federal measures, the NAFTA requires at least 60 days advance notice. Article 718.4 requires a delay between publication of the final sanitary or phytosanitary measure and its effective date. However, there is an exception from these requirements where necessary to address an urgent problem relating to sanitary or phytosanitary protection. These requirements are consistent with requirements already in force for federal agencies under the U.S. Administrative Procedure Act.

Advance notice and publication of sanitary and phytosanitary measures contribute to what is often referred to as the "transparency" of measures. A "transparent process" allows interested persons to know what requirements apply and to adapt their production or other activity to the requirements. Advance notice and comment should also help to avoid problems and provide an opportunity to determine that proposed sanitary and phytosanitary measures conform to NAFTA disciplines.

Article 719 provides additional transparency. It requires each NAFTA government to have an inquiry point that can answer questions from other NAFTA governments and interested persons, and to provide relevant documents regarding the NAFTA country's sanitary and phytosanitary measures.

Article 717 -- Domestic approval. Article 717 provides that a party may require governmental approval for the use of an additive, or establishment of a tolerance for a contaminant, prior to granting access to its domestic market for a food, beverage, or feedstuff containing that additive or contaminant.

Technical cooperation and harmonization. Several NAFTA provisions address technical cooperation and the harmonization of standards. Article 720 provides for technical cooperation between the NAFTA governments. Article 722 establishes a Committee on Sanitary and Phytosanitary Measures whose functions include facilitating the enhancement of food safety and the improvement of sanitary and phytosanitary conditions in the three NAFTA countries, as well as facilitating technical cooperation between the parties in the area of sanitary and phytosanitary measures.

Article 723 provides for technical consultations between the NAFTA governments on matters covered by Chapter Seven's SPS provisions.

Article 723.6 -- Burden of proof. Article 723.6 makes explicit that a NAFTA government asserting that a sanitary or phytosanitary measure of another government is inconsistent with NAFTA has the burden of establishing the inconsistency.

c. Chapter Nine: Standards Related Measures ("SRMs")

Following is a description of the key environmental provisions of Chapter Nine. The Statement of Administration Action provides a more detailed description of this Chapter.

The NAFTA SRMs text explicitly recognizes the right of each NAFTA party to adopt, maintain, or apply any standards-related measure, including measures to enforce SRMs, and to prohibit the import of products until completion of any domestic approval procedure (Article 904). Article 904.2 explicitly recognizes the right of each NAFTA party to establish the levels of safety and protection of human, animal or plant life or health, the environment or consumers it considers appropriate. While there is no obligation under Chapter 9 that a NAFTA party conduct an assessment of risk or base its SRMs on an assessment of risk, it affirms the right of a country to conduct such an assessment in pursuing its legitimate objectives.

Article 909 -- Transparency. Chapter Nine also requires advance public notice and opportunity to comment on proposed SRMs or modifications to SRMs (Article 909). In the case of federal measures, the NAFTA requires at least 60 days prior written public notice. The NAFTA also requires a delay between publication of the final SRM and its effective date. However, there is an exception from these requirements where necessary to address an urgent problem relating to safety or to protection of human, animal or plant life or health, the environment or consumers. These procedures are consistent with U.S. federal agency requirements under the Administrative Procedure Act.

Articles 905, 906 and 913 -- Cooperation and harmonization. Recognizing the crucial role of SRMs in achieving legitimate objectives, the NAFTA parties commit to work jointly to enhance the level of safety and of protection of human, animal and plant life and health, the environment and consumers (Article 906.1). The NAFTA provides for the use of relevant international standards, where effective or appropriate to fulfill the NAFTA party's legitimate objectives, as a basis for each NAFTA party's own SRMs, in order to facilitate trade among the parties. At the same time, the NAFTA explicitly affirms the right of each NAFTA party to adopt or maintain SRMs that achieve a higher level of protection than the relevant international standard (Article 905).

The NAFTA parties also commit to make their respective SRMs compatible to the greatest extent practicable, without reducing the level of safety or of protection of human, animal or plant life or health the environment or consumers (Article 906.2). Greater compatibility should be achieved through the notice and comment procedures outlined above, and through the working groups envisioned under the Agreement.

Other NAFTA SRM provisions include an obligation for each NAFTA party to treat a technical regulation of another NAFTA party as equivalent to its own if the exporting party demonstrates to the importing country's satisfaction that its technical regulation adequately fulfills the importing party's legitimate objectives (Article 906).

Article 913 establishes a Committee on Standards-Related Measures whose functions include facilitating the process by which the countries make their standards-related measures compatible and enhancing cooperation on developing, applying, and enforcing standards-related measures.

Article 912.4 -- Burden of proof. As with Chapter Seven for SPS, Article 912.4 of Chapter Nine provides that in any dispute, the burden of proving any inconsistency of a standards-related measure with NAFTA is on the challenging party.

3. Relationship of NAFTA to International Environmental and Conservation Agreements

The NAFTA gives clear priority to the trade provisions of certain international environmental agreements. During negotiation of the NAFTA, Congress and the environmental community sought to ensure there would be no ambiguity about the relationship between the NAFTA's provisions and the trade provisions of key international environmental agreements. In particular, they requested an explicit assurance that the important trade obligations of the

Montreal Protocol on Substances that Deplete the Ozone Layer, the Convention on International Trade in Endangered Species of Wild Fauna and Flora ("CITES"), and the Basel Convention on the Control of Transboundary Movements of Hazardous Wastes and Their Disposal (or related U.S. bilateral agreements with Canada and Mexico) could be fully implemented without creating any conflicts with NAFTA.

The NAFTA provides that in the event of any inconsistency between NAFTA and the mandatory trade provisions of these environmental agreements, the environmental agreements will prevail to the extent of the inconsistency. Where a party has a choice between equally effective and reasonably available means of complying with its international environmental obligations under these agreements, it should choose the measure that is most consistent with the NAFTA.

NAFTA Article 104 specifically lists the three environmental agreements cited above as agreements whose trade obligations take precedence over any inconsistent obligations under the NAFTA. In addition, the NAFTA provides that the list of international environmental agreements whose trade obligations are to be given precedence can be expanded, by agreement among the NAFTA parties, through listing those agreements in Annex 104.1. Annex 104.1 currently includes the U.S. bilateral agreements with Canada and Mexico regarding transboundary movement of hazardous wastes. The United States has obtained commitments from Canada and Mexico that our two bilateral migratory bird treaties, the Convention on the Protection of Migratory Birds (with Canada) and the Convention Between the United States of America and the United Mexican States for the Protection of Migratory Birds and Game Mammals, will be added to Annex 104.1 once the NAFTA takes effect. The United States is considering which additional international environmental agreements should be added to this list.

4. NAFTA's Dispute Settlement Provisions

NAFTA's dispute settlement provisions include several that are sensitive to environmental concerns. First, as noted above, NAFTA explicitly provides that a NAFTA government challenging another government's environmental measure bears the burden of proving that it is inconsistent with the agreement (Chapter 7, Article 723.6 and Chapter 9, Article 912.4).

Second, in most types of disputes arising under the NAFTA, the dispute settlement panel may, on its own initiative or at the request of a disputing party, request a written report from an independent Scientific Review Board on any issues of fact concerning the environment, health, or safety (Chapter 20, Article 2015). The dispute settlement panel will take the Review Board's report into account before reaching its final decision and will release the report to the public together with any final panel decision that is publicly released (Article 2015). The Commission for Environmental Cooperation, created by the NAFTA environmental side agreement, may also play a role in preventing and resolving NAFTA disputes involving environmental issues (Env. Article 10.6).

Third, if a party to a dispute claims that its action is related to its obligations under one of the international environmental or conservation agreements, or under NAFTA's provisions on standards or sanitary and phytosanitary measures, it has the option of having the dispute considered exclusively under the NAFTA, rather than under the General Agreement on Tariffs and Trade (the "GATT") (Article 2005). Environmental groups requested inclusion of this provision so that the NAFTA environmental provisions would apply in those cases, instead of the less specific provisions of the GATT.

Some objections have been raised regarding the openness of NAFTA's dispute settlement process. In fact, any disputes that may arise under the NAFTA will be between governments, and the United States' primary interest will be in resolving such diplomatic differences. However, the Administration recognizes that the outcome of these disputes may be of great interest to those in the United States outside the government. Accordingly, the Office of the United States Trade Representative will provide, as it has in all recent trade disputes, for public notice and opportunity for input into dispute settlement proceedings involving the United States under the NAFTA.

USTR currently provides public notice of the initiation of trade disputes through publication in the Federal Register. It also briefs interested individuals and groups on dispute settlement proceedings and accepts input from the public into the facts and arguments involved in such proceedings. For example, USTR has met with interested members of the environmental community, industry and congressional staff on numerous occasions to brief them on particular disputes, including the status of the proceeding and the issues involved. USTR also makes available to the public U.S. submissions to dispute settlement panels.

Finally, it should be emphasized that NAFTA dispute settlement results will not supersede U.S. laws unless and until the government acts domestically to implement the results. That will require a public process in the United States (see discussion in Section V.C). The Statement of Administrative Action for Chapter 20 provides a more detailed description of NAFTA's dispute settlement provisions.

5. Investment Chapter Provisions

The NAFTA Investment Chapter (Chapter 11) permits each party to impose stringent environmental requirements to ensure that investment activity in its territory is undertaken in an environmentally sensitive manner, so long as the parties do not discriminate between domestic and foreign investors. This provision permits, for example, the requirement in some states for environmental impact assessments of new private construction, as well as government projects (Article 1114.1).

Further, Article 1114 also addresses one of the key environmental concerns that Congress and the public had about a North American free trade agreement: that businesses might shift their production to a country whose environmental standards were lower, and therefore less costly to comply with (sometimes called the "pollution haven" issue). In Article 1114.2 the parties renounce the relaxation of health, safety or environmental measures for the purpose of attracting or encouraging investment. The text sets forth a procedure for compulsory consultations between parties in case such a relaxation occurs, with the objective of ending the practice.

The Environmental Agreement further addresses "pollution haven" concerns by committing the parties to effective enforcement of their environmental laws, and providing a dispute settlement mechanism to enforce that commitment. (That agreement is discussed in more detail in Section II.B.)

B. SUPPLEMENTAL AGREEMENT ON THE ENVIRONMENT

The North American Agreement on Environmental Cooperation supplements the environmental provisions and objectives of the NAFTA, further ensuring that trade liberalization will not come at the expense of environmental protection. The Environmental Agreement establishes a framework for United States, Mexican and Canadian cooperation on environmental matters and commits the parties to effective enforcement of their environmental laws. The creation of this unprecedented institution and the parties' ability to hold each other accountable for the agreement's environmental commitments hinges, however, upon the entry into force of the NAFTA.

1. Background

The contours of the North American Agreement on Environmental Cooperation ("the Environmental Agreement"), were first laid out by then-candidate Clinton in October of 1992. The agreement was negotiated at the behest of the United States with Mexico and Canada in the spring and summer of 1993 and signed by the respective heads of state on September 14, 1993.

This initiative was launched in response to concerns expressed by President Clinton, as well as by U.S. environmental and conservation groups and the public, over the possible environmental effects of free trade among the three countries. USTR led the negotiations for the Environmental Agreement. Many of the agreement's provisions were developed by an environmental negotiating sub-group co-chaired by the Environmental Protection Agency and the Department of State, and including representatives from the Departments of Agriculture, Commerce, Justice, and Interior. These participating agencies consulted closely with a broad spectrum of business and environmental organizations, and incorporated many of these groups' proposals into U.S. negotiating positions. Many of these positions were ultimately adopted by the parties.

The Environmental Agreement has a broad, inclusive scope. The Agreement establishes a Commission for Environmental Cooperation, which may address any environmental or natural resource issue through its work program. Moreover, any environmental concern or obligation of the Agreement may be the subject of consultations between the parties, from migratory and endangered species to transboundary pollution, to advising the NAFTA Free Trade Commission on disputes on matters related to the environment. Understandably, the realm of issues subject to dispute settlement panels and possible sanctions is more circumscribed, and is focused on whether the parties are effectively enforcing their environmental laws.

The North American Agreement on Environmental Cooperation will enter into force on January 1, 1994, immediately after entry into force of the NAFTA. The Agreement's key provisions are set forth below.

2. Preamble and Objectives

The Preamble sets out the goals, principles and aspirations on which the Agreement is based. It recognizes a tradition of cooperation on the environment, emphasizes the importance of public participation in environmental protection, and expresses a commitment to support and build on international environmental agreements and on existing institutions.

Fundamental objectives of the Agreement, as set forth in Article 1, are cooperation on the conservation, protection and enhancement of the environment of North America and the effective enforcement of and compliance with domestic environmental laws. Other key objectives include the promotion of sustainable development, support for the environmental goals and objectives of the NAFTA, and the promotion of transparency and public participation in the development and improvement of environmental laws and policies.

3. National Obligations

Part Two of the Environmental Agreement spells out a number of obligations the parties have undertaken with respect to their national environmental laws and policies. Many of these obligations -- most notably the commitment to effective environmental law enforcement -- have no precedent in existing trade agreements or international environmental agreements. These obligations apply to subnational levels of government as well as to the federal level.

Ensuring high levels of environmental protection. While affirming the right of each party to establish its own levels of protection, policies, and priorities, and to adopt or modify its laws accordingly, Article 3 of the Agreement requires that each party ensure that its laws provide for high levels of environmental protection and strive to continue to improve those laws. The Agreement also protects the rights of states and provinces to set high levels of protection.

Effective enforcement. To achieve high levels of environmental protection and compliance, each party has committed to effectively enforce its environmental laws through appropriate government actions such as: appointment and training of inspectors; monitoring of compliance and examination of suspected violations of law; seeking voluntary compliance agreements; and using legal proceedings to sanction, or to seek appropriate remedies for, violations of its environmental law (Env. Article 5). A party's failure to observe this obligation may be subject to dispute settlement and sanctions (Part Five). These provisions reinforce the NAFTA Article 1114 provisions designed to protect against countries relaxing environmental protection in order to attract investment.

Transparency and access to enforcement processes. The parties have also agreed to ensure that their procedures for the development and enforcement of environmental laws and regulations are fair, open and equitable (Env. Article 4). Each party has committed to ensure appropriate public access to judicial and administrative procedures for the enforcement of its environmental laws (Env. Articles 5 and 6). Such access for citizens includes the right to request action for the enforcement of domestic environmental law, and to sue another person under that party's jurisdiction for damages.

Other commitments. Each party has also committed:

- to periodically prepare and make publicly available reports on the state of its environment;
- to develop and review environmental emergency preparedness measures;
- to promote environmental education, scientific research, and technological development;
- to assess, as appropriate, environmental impacts; and
- to promote the use of economic instruments for the efficient achievement of environmental goals (Env. Article 2.1).

Each party will also notify the other parties of a decision to ban or severely restrict a pesticide or toxic substance, and will consider banning the export to another party of such substances, the use of which is banned within its own territory (Env. Article 2.2).

4. Commission for Environmental Cooperation

Perhaps the most significant elements of the Environmental Agreement are the provisions creating the trinational Commission for Environmental Cooperation ("CEC"). This new organization will facilitate cooperation among the three NAFTA parties on the full range of environmental and conservation issues, both within the U.S.-Mexico and U.S.-Canada border areas and throughout the territories of the three countries.

The CEC, with its continent-wide geographic scope, is a significant institutional complement to the primarily bilateral institutions that already exist to foster environmental cooperation and avoid or settle environmental disputes between the countries of North America. It provides a structure for the parties to study issues, form working groups, and solve problems of common concern. In addition to the CEC's importance for the region, its form and substance are likely to exert a strong influence on future environmental relationships -- including those relating to trade -- among other countries around the world.

a. Structure of the CEC

Part Three of the environmental agreement sets forth the structure and functions of the CEC. A governing council of the environmental ministers from Mexico, Canada and the United States will oversee the Commissions's day-to-day work (Articles 9 and 10). That work will be carried out by a permanent, independent Secretariat headed by an Executive Director (Article 11), with the advice of a trinational Joint Public Advisory Committee consisting of five non-governmental advisors from each country (Article 16). The functions of each of these three components is described more fully below.

b. Important aspects of the CEC

The Council of Ministers. The Council of Ministers, supported by a full-time, permanent and independent Secretariat, will serve as a forum for discussing and making recommendations on all issues, including life-cycle management, transboundary environmental harm, natural resource accounting methods and ecosystem protection. The Council will also serve as a forum for discussing and settling actual or potential environmental disputes. The U.S. representative on the Council will be the Administrator of the Environmental Protection Agency. Close coordination among interested U.S. government agencies, including those with responsibility for conservation and environmental protection, will be necessary.

Meetings of the Council of Ministers. The Council will meet in regular session at least annually and in special session at any time at the request of any party. All annual meetings will have public sessions and the Council may open any other meetings to the public as well. Except as specifically provided, all decisions and recommendations of the Council are to be made public.

Council agenda items. The Council may take up and develop recommendations on any environmental matter. The parties to the Environmental Agreement have agreed to highlight certain areas for discussion and development of recommendations. These include a number of environmental issues to which the United States intends to give particularly high priority, such as: (a) comparability of techniques and methodologies for data gathering and analysis; (b) pollution prevention techniques and strategies; (c) transboundary and border environmental issues; and, most notably, (d) the environmental implications of goods throughout their life cycles (including processes and production methods and disposal techniques). Work on this latter issue will respond to significant public concerns about how to address the environmental impacts of processes and production methods, and will also address the importance of internalizing environmental costs for both trade and the environment.

The Agreement gives special emphasis to:

- Strengthening cooperation on the development and continuing improvement of environmental laws and regulations (Article 10.3), and encouraging effective environmental enforcement and compliance, and technical cooperation (Article 10.4).

- Promotion and development of recommendations on granting residents of the three countries greater access to environmental information held by public authorities, including information on hazardous materials in communities ("community right-to-know") (Env. Article 10.5.a). The United States will promote full public disclosure by the parties of information collected under the U.S.-Mexico hazardous waste tracking system. The United States also will seek to assure that all parties adopt and strengthen community right-to-know requirements for toxic releases as expeditiously as possible.

- Consideration and development of recommendations for assessing the environmental impacts of proposed government projects likely to cause significant adverse transboundary effects, notification and provision of information concerning such projects, and mitigation of potential adverse environmental effects. The parties are to endeavor to reach agreement on this item within three years (Env. Article 10.7).

- Encouragement of each NAFTA party to establish administrative procedures permitting the other parties to seek the reciprocal reduction and elimination or mitigation of transboundary pollution; and the consideration and possible development of recommendations regarding parties' provision of reciprocal rights and remedies before their courts and administrative agencies for persons in other NAFTA parties suffering damage or injury caused by pollution originating in their territory (Env. Article 10.8 and 10.9).

Independent Secretariat. The Secretariat's Executive Director will be chosen by the Council for a single, renewable three-year term and is removable solely for cause. The Executive Director will select the Secretariat staff, subject only to a two-thirds veto by the Council of any individual chosen who does not meet the standards set in accordance with the Agreement. The Executive Director will consider potential candidates put forward by the parties and the Joint Public Advisory Committee, strive for proportional representation, and make choices based strictly on efficiency, competence and integrity. The Secretariat will develop the Commission's annual program and budget for approval by the Council. Each budget will provide for contingencies. This means that funds will be available to report on time-sensitive matters brought to the Secretariat's attention by the public, unless the Council acts affirmatively by a two-thirds vote to stop such an investigation.

Secretariat reports. A principal reporting function of the Secretariat involves preparation of an annual report for release to the public covering the Commission's activities during the previous year (Env. Article 12). This report will include recommendations on issues that it has reviewed or investigated. The Secretariat will also report on actions taken by the parties with respect to their obligations under the Environmental Agreement, including collection of data on enforcement activities and information submitted by the public.

In addition to this annual reporting function, the Secretariat:

- may report publicly on any matter within the scope of the annual program or any other matter (not related to the failure to enforce environmental laws) brought to its attention through public submissions or otherwise, of which the Council does not affirmatively disapprove (Env. Article 13);

- may, upon a two-thirds vote of the Council, develop a factual record based upon a submission from the public that a party is failing to effectively enforce its environmental law, provided the submission containing the assertion meets certain basic threshold requirements. These factual reports will be made publicly available upon a two-thirds vote of the Council (Env. Articles 14 and 15);

- will periodically report on the state of the environment of the parties (Env. Article 12.3); and

- will assist the Council in conducting an ongoing consideration of the environmental effects of the implementation of the NAFTA, as tariff and non-tariff barriers are eliminated over a 15-year period (Env. Article 10.6).

Secretariat's access to information. In preparing any report, the Secretariat may draw not only upon public information but also upon information submitted by the parties, the Joint Public Advisory Committees, or interested non-governmental organizations and persons, gathered through public consultations, or otherwise developed by the Secretariat or by independent experts (Env. Article 13). Each party must provide such information as the Secretariat or Council requests, provided, however, that whenever domestic law does not permit release of the information or for some other reason the party does not make the information available, the party must promptly notify the Council of its reasons in writing (Env. Article 21). Information submitted in confidence will be kept confidential (Env. Article 11.8).

Joint Public Advisory Committee. The Joint Public Advisory Committee will include five members of the public from each country. It will meet at least once a year, concurrent with the regular session of the Council. The Joint Committee will advise the Council and provide technical, scientific or other information to the Secretariat. It will also provide input to the annual program and budget of the Council as well as to the annual and other reports.

Relationship of the CEC to NAFTA institutions. It is intended that the institutions created under the NAFTA and the Environmental Agreement will support and augment one another. One of the principal objectives of the Environmental Agreement is to support the environmental goals and objectives of the NAFTA. This will be achieved primarily through cooperation between the CEC and the NAFTA Free Trade Commission. Such cooperation will take the following forms, as spelled out in Article 10.6 of the Environmental Agreement:

- The CEC will act as a point of public inquiry and a place for receipt of public comments concerning NAFTA's environmental goals and objectives.

- Where there are consultations under the NAFTA because one party believes another to have derogated from or waived an environmental measure for the purpose of attracting investment, the CEC may be called upon to provide expert assistance.

- The CEC will work to prevent environment-related trade disputes or resolve them as they arise, by seeking to avoid disputes among the parties, by making recommendations to the Free Trade Commission as to how such disputes may be avoided, and by identifying experts capable of providing needed technical and scientific advice to NAFTA committees and working groups.

- The CEC will consider on an ongoing basis the environmental effects of NAFTA implementation and otherwise assist the Free Trade Commission in environment-related matters.

CEC role in promoting effective enforcement. The Commission will promote and contribute to improved environmental enforcement throughout North America. It will encourage the sharing of environmental enforcement technologies and enforcement information between the parties, and will report on their environmental enforcement activities (Env. Articles 10 and 12). As described above, the Secretariat will prepare factual records on enforcement matters based on submissions from the public in the three countries. And, as described more fully below, the Council will address complaints between parties regarding compliance with the obligation to effectively enforce environmental laws, resolving them through consultations or the establishment of dispute settlement panels.

5. Consultations

A party may request consultations regarding any matter that affects the operation of the Agreement. Should the consultations fail to resolve the matter, any party may call a meeting of the Council. In seeking a resolution, the Council may consult technical advisors or create working groups or expert groups and make recommendations (Articles 10.1, 9.5, 22, and 23).

6. Resolution of Disputes Regarding Enforcement of Environmental Laws

The Environmental Agreement establishes a dispute settlement mechanism to ensure that the parties effectively enforce their environmental laws. The primary objective of the dispute settlement process is to correct problems of nonenforcement, not to mete out punitive measures. Accordingly, parties found to have engaged in a persistent pattern of failure to effectively enforce their laws are required to correct the problem in the first instance; this is then followed by the imposition of monetary fines for failure to take appropriate corrective steps; and, as a last resort, trade sanctions may be imposed (or, in the case of Canada, the panel's decision may be enforced against the government in Canadian court).

Formation of a dispute resolution panel. Any party may request the formation of an arbitral panel if the Council has not succeeded in resolving a dispute involving a party's alleged persistent pattern of failure to effectively enforce an environmental law that relates to a situation involving workplaces, firms, companies or sectors that produce or compete with goods or services traded between the parties. A panel will be established on a two-thirds vote of the Council (Env. Article 24).

Panelists will normally be chosen from a previously agreed roster of objective, independent experts, including experts on environmental matters (Env. Article 25.2). The United States will seek to assure that Panel members have significant environmental experience and expertise. With the approval of the disputing parties, a panel may seek information and technical advice from any person or body that it deems appropriate (Env. Article 30). The final report of the panel will be made publicly available five days after it is transmitted to the Council.

Implementation of a panel report. If a panel makes a finding that a party has engaged in a persistent pattern of failure to effectively enforce its environmental law, the parties may, within 60 days, agree on a mutually satisfactory "action plan" to remedy the non-enforcement (Env. Articles 33 and 34).

If there is no agreed action plan, then between 60 and 120 days after the final panel report, the panel may be reconvened to evaluate an action plan proposed by the party complained against or to set out an action plan in its stead. The panel may also make a determination on the imposition of monetary enforcement assessments on the Party complained against (Env. Article 34).

The panel may be reconvened at any time to determine whether an action plan is being fully implemented. If it is not being fully implemented, the panel is to impose a monetary enforcement assessment on the party complained against (Env. Articles 34 and 35; Annex 34). The assessment may be up to $ U.S. 20 million for the first year after entry into force of the agreement, thereafter it may not exceed 0.007% of total annual trade in goods between the parties.

In the event that a party complained against fails to pay a monetary enforcement assessment or continues in its failure to enforce its environmental law, the party is liable to ongoing enforcement actions. In the case of Canada, the Commission, on the request of a complaining party will collect the monetary enforcement assessment and enforce an action plan in summary proceedings before a Canadian court of competent jurisdiction (Env. Annex 36A). In the case of Mexico and the United States, the complaining party or parties may suspend NAFTA benefits based on the amount of the assessment (Env. Article 36; Annex 36B).

7. Transparency and Public Participation

The United States is committed to ensure the transparency and openness of activities under the Environmental Agreement. In general, the United States will support making available to the public all non-confidential elements of reports, factual records, decisions, recommendations, and other information gathered or prepared by the Secretariat or Council. In cases where written information is not made available, the United States will call for written explanations setting forth the reasons for the decision. The Administration also is committed to maintaining close consultation with the states and local authorities in carrying out activities under the Environmental Agreement.

In establishing the Model Rules of Procedure for dispute settlement (Env. Article 28), the United States will seek to assure public access to panel proceedings, written submissions and panel reports, and develop other mechanisms for public access to and involvement in the process.

C. U.S.-MEXICO BORDER ENVIRONMENT COOPERATION AGREEMENT

Although a number of agreements and institutions are in place that address border environmental issues, the United States and Mexico have recognized that a new binational mechanism is needed to strengthen environmental cooperation among interested parties in the border region. Accordingly, they have reached an agreement to establish two new institutions: a Border Environment Cooperation Commission ("BECC") to help coordinate and develop environmental infrastructure projects; and a North American Development Bank ("NADBank") to help generate the financial resources for constructing the projects.

1. Background

The U.S.-Mexico border region has serious environmental problems that must be addressed. Many of these problems stem from transboundary movement of pollutants, while others result from inadequate funding for basic services such as a clean water supply, wastewater treatment, and facilities for the sound management of solid waste. To control transboundary pollution and to capture economies of scale for neighboring communities, border environmental concerns can be most effectively addressed through joint action.

The U.S.-Mexico Border Environment Corporation Agreement represents a significant additional commitment by Mexico and the United States to implement effective solutions to environmental problems in the border region. It provides mechanisms for the two governments to support a wide range of environmental projects in the border region, and even certain projects outside that region that the governments determine have significant transboundary environmental effects. Initially, preference will be given to projects addressing the most pressing environmental and public health needs, such as wastewater treatment, clean drinking water supply, and management of solid waste.

This agreement augments the NAFTA by helping to ensure that the environmental consequences for the border area of increased trade with Mexico will be affirmatively managed to secure a better environment for the millions of Americans who live there, as well as their Mexican neighbors. As with the Environmental Agreement, this new agreement will come into force only upon entry into force of the NAFTA.

2. The Border Environment Cooperation Commission

The key to the operation of the new agreement is the BECC. The BECC will work with the affected states and local communities and non-governmental organizations to help them formulate effective solutions to environmental problems in the border region. It will also certify projects as eligible for financing through the NADBank.

The BECC will work with state agencies, local communities, and other project sponsors in developing and implementing environmental infrastructure projects. The BECC will not develop or manage projects itself. Rather, the parties envision that it will assist the border communities by providing technical, environmental, and financial expertise to all phases of a project.

The BECC will help local communities and private parties to coordinate projects so as to provide the most effective and efficient solution for the environmental needs of the area. This function is especially important to ensure coordination of projects across the international boundary, so that Sister Cities can work together to solve shared problems. Through a combination of its own staff, engineering staff of the International Boundary and Water Commission, and private contractors, the BECC can also provide a full range of project services, including engineering, design, project siting, environmental analysis, and oversight of construction and operation.

The BECC will also have financial expertise to evaluate the financial feasibility of projects, including the level of user fees required to service project revenue bonds. The BECC can help project sponsors to arrange public and private financing for projects. In some cases, financing may be available directly through private markets or in combination with government funds.

For NADBank financing to be used, the BECC must certify that the project meets appropriate technical, financial, and environmental criteria. The BECC may establish both general guidelines and project-specific criteria for certification of projects. The agreement affirms that each project must meet all environmental requirements for the place where the project will be located or carried out, including applicable requirements for environmental impact assessments.

For a project that would have significant transboundary effects to be eligible for certification, an environmental assessment must be prepared, and the Board of the BECC must determine, in consultation with affected states and localities, that the project meets the conditions necessary to achieve a high level of environmental protection for the affected area.

The BECC will have a binational Board of Directors. Two members of the Board for each country will be federal officials, including the Administrator of EPA and the IBWC Commissioner, but the other three members from each country will be drawn from the border area, representing state, local, and public interests. The Board must consult on major issues with an Advisory Council, also drawn predominantly from the border area. Non-governmental organizations, including community, business and environmental associations, are eligible for representation on both the Board and the Council. The Advisory Council will have a consulting role with respect to general guidelines, environmental criteria applied to projects, and other aspects of the certification process. Finally, the agreement provides expressly for public notice and opportunity to comment before important decisions. This governance structure will ensure that the views of affected states, local communities, and the general public will be fully taken into account in the work of the Commission.

3. The North American Development Bank

The North American Development Bank will be capitalized and governed equally by the two countries. Its primary purpose is to finance projects certified by the Border Environment

Cooperation Commission. Based on its paid-in capitalization, the parties envision that the NADBank will be able to make available some $2 billion or more in loans and guarantees for environmental infrastructure projects, with an upper limit of just under $3 billion.

The NADBank will use its own capital funds (contributed equally by the United States and Mexico), funds raised by it in the financial markets, and other available resources to finance public and private investment in environmental infrastructure projects, and to encourage and supplement private investment in environmental infrastructure projects. Initial paid-in capital will be $450 million, comprised of $225 million each from Mexico and the United States. The callable capital will amount to $2.55 billion, also contributed equally by the United States and Mexico.

Each country may apply up to ten percent of its capital to a community adjustment and investment "window," to supplement existing government assistance programs.

The NADBank is intended to supplement existing sources of financing. It will support, not impair, the ability of governments and investors to seek financing from other institutions. In conjunction with other sources of funds, such as existing governmental sources of funds in the United States for U.S. projects, as well as World Bank and Inter-American Development Bank loans to Mexico, a total of $7-8 billion of financing is expected to be available for environmental improvements in the border area over the next decade.

III. MEXICO'S POLLUTION CONTROL REGIME

A. THE GENERAL ECOLOGY LAW

In order to address "pollution haven" concerns, in 1993 the U.S. EPA carried out a comparison of U.S. and Mexican environmental laws. This comparison covered water, air, hazardous waste, pesticides, and industrial chemicals, among other topics. As a result of this study, EPA has concluded that overall (with certain exceptions), the United States and Mexican regulatory regimes are designed to achieve comparable levels of environmental protection.

A noteworthy exception is to be found in the case of Mexico's significantly higher sulfur dioxide source-specific (e.g., coal-fired power plant) emissions standards. The United States most recently tightened its controls on sulfur dioxide from existing power plants in the 1990 amendments to the Clean Air Act.

With regard to pesticides, Mexico usually follows U.S. residue tolerances. Otherwise, it follows tolerances set by an international standards organization known as Codex Alimentarius or other developed countries. Only a few pesticides unregistered in the United States are registered in Mexico, such as DDT and BHC (neither of which are registered for food uses in Mexico).

1. Background

Since the announcement in September 1990 that NAFTA would be negotiated, concerns have been expressed that Mexico's environmental laws, regulations and standards could, in certain instances, provide lower levels of environmental protection than U.S. environmental laws, regulations and standards. Concerns have also been expressed that, in a free trade area created under NAFTA, Mexico could become a haven for industries seeking to avoid higher environmental compliance costs in the United States. The results of this scenario, according to proponents of the "pollution haven" argument, would be: (1) decreased U.S. competitiveness, as U.S. industries move manufacturing operations to Mexico to take advantage of lower environmental compliance costs; (2) an increase in industrial pollution in Mexico, due to the larger number of polluting industries taking advantage of Mexico's supposedly lower environmental standards; (3) a rise in the degree and amount of transboundary pollution crossing into the United States from industries located in Mexico; and (4) the eventual lowering of U.S. environmental standards to stem the loss of industry and jobs from the United States.

As discussed in Sections II.A and II.B, NAFTA and the Environmental Agreement directly address pollution haven concerns through provisions that encourage the upward harmonization of environmental standards, strengthen enforcement of those standards, and enjoin governments from relaxing standards for the purposes of attracting investment.

In response to "pollution haven" concerns, as the NAFTA negotiations commenced, EPA began to gather information about Mexican environmental laws from Mexican officials. The process began with an examination of Mexico's General Ecology Law of 1988, a multi-media "umbrella" statute. EPA also examined media-specific regulations promulgated under the General Ecology Law for such areas as air, water, waste, and environmental impact assessment, as well as the limited number of environmental standards which were associated with the regulations at that stage in the development of Mexican environmental law. Based on information gathered from meetings in Mexico, EPA publicly released an interim report on Mexican environmental law in November 1991.

On the issue of comparability of U.S. environmental law and Mexican environmental law, the interim report concluded:

> " . . . Mexico's environmental laws, regulations and standards are in many respects similar to those in the United States. The 1988 General Law of Ecological Balance and Environmental Protection ("General Ecology Law") embodies many principles and approaches similar to ours. . . . The regulations and technical standards implementing the Mexican law are generally comparable to their counterparts in the United States, although each regime includes provisions that the other lacks. To the extent that differences in scope are due to the early stage of development of Mexico's program, it would be premature to draw too many conclusions about overall stringency or comparability."

The process of understanding Mexico's evolving environmental law regime, including the official reorganization of Mexico's environment secretariat in June 1992, continued when EPA and officials of Mexico's Secretariat of Social Development ("SEDESOL") met to discuss legal developments in March 1993.

Since March 1992, Mexico has been in the process (not yet completed) of reissuing all of its existing 83 environmental standards, after having subjected each to a cost-benefit analysis. By the end of 1994, Mexico will have finished a process, already well underway, of releasing 125 new, additional environmental standards which will have been subjected to the same cost-benefit analysis. Following the issuance of these standards, Mexico will continue to develop its regulatory regime, in order to address a greater range of environmental problems. It is important to remember that Mexico's General Ecology Law is only five years old and that Mexico has had only five years to construct an entire environmental regulatory regime.

In the summer of 1993, using information it has obtained to date, EPA initiated a comparison of U.S. and Mexican environmental standards in a number of industrial sectors in each of four principal media areas: water, air, hazardous waste, and pesticides and industrial chemicals.

With certain important exceptions where standards diverge significantly (as with sulfur dioxide emissions controls for coal-fired power plants), EPA has concluded that many of the Mexican environmental standards in the principal media areas are broadly comparable to U.S. standards and that, overall, the two regulatory regimes are designed to achieve comparable levels of environmental protection. EPA is now in the process of completing its findings, which will be made available to Congress and the public.

Following is a preliminary summary of EPA's findings to date in these media areas.

2. Water

The scope of Mexico's 1988 General Ecology Law appears fairly comprehensive with respect to water pollution, covering: releases from industry, municipalities, agriculture and livestock activities, and mining; use of pesticides, fertilizers, and toxic substances; infiltrations into aquifers; solid waste dumping; and federal facilities. The law prescribes principles for developing water quality and other technical standards, as well as for exploiting and conserving marine resources.

The Mexican water pollution legal regime generally appears to contemplate a regulatory system that would control both point and non-point sources of pollution as broadly as the U.S. Clean Water Act. In fact, the Mexican system appears to go beyond its U.S. counterpart in applying to discharges into groundwater as well. Both Mexico's General Ecology Law and the U.S. Clean Water Act provide for the adoption of wastewater discharge restrictions implemented through a federal/state permitting program. Both statutes rely on technology-based controls and effluent limitations, water quality standards and consideration of the assimilative capacity of the receiving waters in determining the level of control that is necessary for a given source.

Mexico's requirements for permitting point source discharges and for setting maximum permissible limits appear to be comparable to the Clean Water Act's permit and discharge limitation system. Its water quality criteria, which appear to be similar to EPA's water quality criteria, are based strictly on scientific evidence. Unlike EPA's water quality criteria, however, Mexican water quality criteria do not appear to form the basis for discharge conditions. In the United States, a use is first designated, and then non-binding criteria are used to develop water quality standards to help ensure attainment of that use. The Mexican requirements for wastewater treatment, protection of watersheds and prevention of interference with municipal wastewater treatment systems also seem to be parallel to the U.S. system.

Like effluent limitation guidelines and standards promulgated by EPA, the Mexican standards are based on economic and technical feasibility and often reflect technologies of sedimentation, flocculation and precipitation. A significant difference, however, is that the Mexican standards appear to focus primarily on the control of conventional pollutants rather than toxic pollutants.

EPA conducted a preliminary examination of Mexico's direct discharge standards for five industries (petroleum refining, iron and steel production, copper formation, wood preserving, and metal finishing) which have significant operations in the Mexican border area near the United States. That examination revealed the following:

- In general, the Mexican standards for all five industries tend to be in the same concentration range as the U.S. effluent guidelines and standards for the least stringent level of pollution reduction for direct dischargers (i.e., the "best practicable control technology" currently available).

- Considering that formal Mexican water pollution control efforts only date back to 1988, it appears from its published standards that Mexico has laid the groundwork for a meaningful pollution control effort for industrial wastewater discharges to surface waters. However, the Mexican standard for wood preserving is not as stringent as the U.S standard, which is zero discharge for most of the industry.

It is important to note that Mexico's control system regarding municipal wastewater treatment facilities is not yet fully developed. In the United States, municipal treatment systems must comply with secondary treatment requirements unless a special waiver is granted by EPA. These systems must receive a permit for effluent discharges. In Mexico, sources that discharge into municipal treatment systems are subject to federal pretreatment of indirect discharges and standards for municipal systems, and a technical ecological norm establishes discharges into such systems. However, there does not appear to be any federal requirement in Mexico that local municipal systems meet secondary, or even primary, treatment requirements, as those requirements are defined in the United States.

Based on the legal requirements alone, Mexico and the United States would appear to have generally comparable regimes to control water pollution. The extent to which the two systems may differ in practice depends largely on how the maximum permissible limits, criteria, permitting system and other requirements are implemented.

3. Air

Both the Mexican and U.S. air pollution control programs require adoption of ambient air quality standards for certain specific pollutants. Mexico has issued such standards, called "maximum permissible levles" ("MPLs"), for ozóne, carbon monoxide (CO), sulfur dioxide (SO_2), nitrogen oxide (NO_x), total suspended particulates (TSP), and lead. With the exception of particulate matter, which is now regulated as "PM-10" (particulate matter under 10 microns) in the United States, these are the same pollutants covered by the U.S. National Ambient Air Quality Standards ("NAAQS"). Moreover, all of the Mexican MPLs are set at the same level or nearly the same level as the equivalent U.S. NAAQS. (However, Mexico does not have standards to protect the public welfare, which the United States refers to as secondary NAAQS.)

For example, both the United States and Mexico have ambient sulfur dioxide emissions standards intended to prevent SO_2 emissions from "loading" a particular area's air beyond a limit selected to be protective of human health and to prevent other environmental damage. The Mexican ambient health standard for SO_2 is 0.13 parts per million (ppm) averaged over a twenty-four hour period. The primary U.S. national ambient air quality standard for SO_2 is 0.14 ppm averaged over a twenty-four hour period. The Mexican ambient SO_2 emissions standard is therefore slightly more stringent than the equivalent U.S. ambient SO_2 emissions standard.

Mexico and the United States have different approaches for attaining ambient standards. In the United States, the states are responsible, with federal oversight, for assuring NAAQS attainment. The states develop State Implementation Plans ("SIPs"), which are submitted to EPA for approval. Attainment and maintenance of the standards are demonstrated through air quality modeling, which relates emissions to ambient air quality standards. Mexico, on the other hand, relies on a source permitting program instead of on state or local air quality planning with federal oversight.

Like the United States, Mexico has developed a system for further restricting emissions in chronically polluted and vulnerable areas, called "critical zones." A critical zone in Mexico is defined as a zone in which a high concentration of contaminants is found due to topographic and meteorological conditions. Two critical zones along the U.S.-Mexico border, one for Tijuana, and one for Ciudad Juarez, have been designated because of local air pollution problems.

The General Ecology Law also authorizes SEDESOL to engage in a certain amount of additional land use planning to protect some areas and control others. Article 115 of the General Ecology Law states that when SEDESOL determines land uses under urban development programs, consideration should be given to topographical and meteorological conditions to ensure proper dispersion of pollutants. The U.S. Clean Air Act provides a similar, though less sweeping, provision in non-attainment areas.

Mexico controls stationary source air emissions through a source registration and permitting program. SEDESOL has a standard application form for obtaining an operating license, required under Articles 18 and 19 of Mexico's national air regulation. Article 19 specifies the required information, which appears to be similar to what EPA is requiring under the Clean Air Act operating permit regulations.

Mexico's technical norms to control stationary source air emissions resemble U.S. new source performance standards in that they set maximum permissible emissions levels for various pollutants per unit measure of raw material or production. The Mexican norms, however, apply to both new and existing sources, while the U.S. standards apply only to new sources.

Preliminary analysis of Mexico's source-specific emissions standards for specific industries indicates that there is a wide divergence between some U.S. standards and the equivalent Mexican standards. For example, the Mexican SO_2 emissions standard for coal-fired power plants allows a much higher emission rate than EPA's new source programs.[2] Discrepancies have been noted with respect to petroleum-fired plants as well.

However, at this time, it is difficult to reach firm conclusions regarding the comparative stringency of the norms applicable to stationary sources, since Mexico is still in the process of phasing out its existing technical norms and adopting new standards. Study in the area of source-specific standards for stationary sources of air pollutant emissions is continuing.

Like the United States, Mexico has a requirement for reporting emissions data to the government. Once a source has a permit, it must submit annual reports, which include stack

---

[2] An example of this discrepancy which has received significant attention is the Mexican pollution control requirements for two coal-fired power plants constructed along the U.S.-Mexican border, the Carbon I and II facilities. This situation is discussed in greater detail in Section VI.B.

test data. The permit must be modified if changes are made in the source. In the absence of modification, the lifetime of a permit is unclear. SEDESOL reviews the submitted data and, if a violation appears to have occurred, may inspect the source and close it partially or completely, temporarily or permanently, or may impose a fine.

Regarding hazardous air pollutants, Mexico's law seems not to contain any program comparable to that established by the Clean Air Act Amendments of 1990, although it does appear to authorize development of such standards. Mexican law addresses toxic emissions by requiring prior authorization for the emission of hazardous air pollutants. Moreover, Mexico may adopt technical norms in the future which would regulate toxic emissions.

Mexico's mobile source control program is more easily compared to its U.S. counterpart, since both rely on comparable approaches, such as tailpipe emission standards, vehicle inspection and maintenance, fuel content requirements and transportation controls. The differences between the U.S. and Mexican systems lie primarily in the extent to which each of these approaches is used, and in the stage of development of the various implementing programs.

Mexico appears to be moving quickly toward establishing a tailpipe emission control program which is comparable to the one required in the United States. The original technical norm for light-duty motor vehicles requires decreasing emissions beginning in 1989, with dramatic reductions beginning in 1991, apparently envisioning a phasing-in of cars with catalytic converters. The 1993 maximum permissible emissions levels for hydrocarbons, carbon monoxide, and nitrogen oxides are comparable to current U.S. tailpipe standards under the Clean Air Act. Mexico has reissued many of its original standards for internal combustion engines and added standards for various classes of diesel fuels.

Inspection and maintenance of vehicles are critical to the success of any emissions control program. Mexico plans such a program, although it is still being developed in most parts of the country.

The General Ecology Law and Mexico's air regulations provide for development and implementation of transportation controls. Restrictions on automobile use, though rarely adopted in the United States, are fairly severe in Mexico City, and have been a matter of everyday life there for two years. Each car is prohibited from being driven one day out of the five-day work week. In addition, driving may be suspended in certain parts of the city when ambient pollution levels are high.

4. Hazardous Waste

Mexico's General Ecology Law, like the U.S. Resource Conservation and Recovery Act ("RCRA"), strives to regulate activities dealing with hazardous waste, from generation, storage, treatment, and transportation, to final disposition. Article 151 of the General Ecology Law parallels U.S. law by requiring government authorization prior to the initiation of hazardous waste management activities.

Mexican criteria for determining what constitutes a hazardous waste are very similar to U.S. criteria. However, Mexico includes one criterion, "explosiveness," that is not used in the

U.S. program. It is noteworthy that 23 of the 27 chemicals on the Mexican hazardous waste list that are also considered hazardous under RCRA have maximum permissible concentration levels that are lower than their U.S. toxic chemical leaching procedure equivalent.

Mexican regulations provide considerable detail on determining incompatibility between two or more hazardous wastes to prevent problems that can occur during improper storage. While EPA prohibits the storage of incompatible hazardous wastes, it has not promulgated regulations explaining how incompatibility is to be determined.

Under Mexican law, persons who wish to construct a facility that will either generate or manage hazardous waste must receive prior authorization from SEDESOL's National Institute for Ecology ("INE"). The authorization application is similar to that required under RCRA, which requires both general information about the proposed facility and extensive highly technical information.

Furthermore, when dealing with high-risk activities, the applicant in Mexico must prepare a risk study. This risk study deals primarily with the dangers involved in high risk activities, such as the implications of an explosion at the facility site or the release of hazardous waste. In comparison, RCRA does not require environmental impact and risk studies in considering siting of a facility.

Construction of new facilities is subject to detailed location standards,which set forth the requirements that a site must meet if it is to be used for the controlled "confinement" of hazardous wastes. In some respects (e.g., siting landfills in zones connected to aquifers), these standards are more stringent than their U.S. counterparts; in others (e.g., siting in flood and seismic zones), the Mexican approach is less stringent.

Few off-site waste disposal facilities are currently authorized and operating in Mexico. SEDESOL officials recognize the need to develop more waste disposal capacity but note that efforts to develop that capacity may be hampered by Mexican state laws which can validly prohibit the importation of hazardous waste. However, SEDESOL is promoting the creation of waste management facilities. As of the beginning of 1993, there were 22 applications for the creation of these new waste management facilities, with eight applications projected to be approved by INE by the end of February of 1993.

Hazardous waste in Mexico must ultimately be disposed of in a controlled confinement or disposal facility in accordance with applicable ecological technical norms and other requirements. Two significant differences between the U.S. and Mexican legal regimes are that SEDESOL has not yet issued treatment-oriented land disposal restrictions comparable to those under RCRA or addressed the issue of leaking underground storage tanks. SEDESOL has indicated, however, that it intends to address these issues in the near future and currently has the authority to do so.

Mexican law on manifesting hazardous waste appears similar to its U.S. counterpart since manifests are required for the delivery, transport, and reception of hazardous waste, as well as for any "incidents" involving hazardous waste. Manifests must also be submitted to SEDESOL with each shipment, and disposal facilities must file monthly and biannual reports.

Furthermore, the Mexican law appears to require more oversight over the generation of hazardous waste than U.S. law. Unlike the United States, Mexico requires new generators to receive prior authorization by SEDESOL. Mexican law also requires the facilities to periodically report the volume and type of hazardous waste that is generated. In the United States, RCRA authorizations are not applicable until the waste is treated, stored (for more than 90 days), or disposed of. Mexico also requires new and existing facilities to reduce or minimize the volume of waste generated and then apply physical, chemical, or biological treatment to the waste.

Releases of hazardous constituents are not allowed under Mexican law. For active hazardous waste management facilities, both the U.S. and Mexican programs may require corrective action for releases. RCRA requires that owners and operators of hazardous waste facilities undertake corrective action for release of hazardous constituents. In Mexico, this requirement is implemented by means of a given facility's operating authorization.

Although Mexico does not have an equivalent to the U.S. Superfund law or RCRA "corrective action" program to address releases from inactive sites, it does have a fledgling voluntary program. Through contributions being solicited from industry, SEDESOL hopes to build a fund that will help to provide for the cleanup of abandoned hazardous waste sites. To date, approximately 20 enterprises have either contributed or expressed a willingness to contribute to this fund. SEDESOL's role will be to: (1) identify sites; (2) select remedial action; and (3) provide oversight. Industry will be responsible for hiring the contractor and undertaking actual cleanup operations. To date, no systematic effort has been made to identify all the sites where cleanup is needed. However, it is likely that Mexico faces a huge problem in remedying existing hazardous waste contamination. At its current size and rate of growth, the "voluntary fund" may be inadequate to support comprehensive cleanup operations.

While Mexico's hazardous waste standards are not identical to EPA's regulations, they are structured in a similar manner. There are, however, a number of significant differences that may affect the overall stringency of the two programs. Most importantly, unlike the U.S. regime, the Mexican scheme does not impose a general ban on the land disposal of untreated hazardous waste, although SEDESOL officials have indicated an intention to move toward a program like EPA's land disposal restriction program. In addition, while Mexico's regulations do require leachate collection and treatment, they do not require the installation of a double liner below the waste deposited. (The double liner is required by U.S. law for all landfills constructed after 1985.)

RCRA contains extensive requirements relating to groundwater monitoring, closure, and a facility's financial ability to provide for closure and clean-up. The Mexican law does not appear to impose any such financial responsibility requirements and provides very little detail on closure or groundwater monitoring requirements. And, while the requirements for "confinement cells" are quite detailed, there do not appear to be specific requirements for other types of units, such as tanks and incinerators. Finally, Mexico's regulations do not appear yet to apply to underground storage tanks.

5. Pesticides

Mexican officials often adopt U.S pesticide residue tolerances as the official Mexican tolerances, where such U.S. tolerances exist. Otherwise, limits set by an international standards organization known as Codex Alimentarius, or the limits of other developed countries, are usually adopted.

Only a few pesticides that are banned or unregistered in the United States are registered in Mexico. DDT is registered for use in malaria control programs by public health officials, a use approved by the World Health Organization. BHC is authorized as a restricted-use pesticide for locust control. However, neither of these two pesticides is registered for food uses in Mexico. In addition, Mexico is considering banning EPN and chlordane, which are prohibited for food uses in the United States.

Mexico's data requirements for the registration of pesticides are almost identical to those used by EPA. Both the United States and Mexico require toxicological data, efficacy data, and long-term environmental effects studies. However, since almost all pesticides used in Mexico are imported, the registration authorities do not review every individual study, relying instead on the review in a developed country of origin which has approved the pesticide. In addition to these data, a certificate of registration in the country of origin is required. If Mexico determines that the foreign data do not account for Mexican weather conditions or climate, then additional data that take these factors into consideration is required.

While there are no specific requirements in Mexico for good laboratory practices to be followed in generating pesticide data, most countries that supply pesticides to Mexico subscribe to good laboratory practices as members of the OECD.

Mexico's pesticide labelling practices are consistent with those of the United States and with the U.N. Food and Agriculture Organization ("FAO") Guidelines on Good Labelling Practice. Mexico's labelling requirements include: requirements for ingredient statements; toxicity category; use directions (including crops on which the pesticide may be applied); as well as warnings and cautions.

Unlike the United States, pesticides whose use is prohibited in Mexico cannot be manufactured there. (In the Environmental Agreement the United States has agreed to consider banning the export to NAFTA countries of pesticides whose use is prohibited domestically.)

There are three categories of pesticide tolerance differences between the United States and Mexico:

- fifty-eight pesticides have tolerances in both countries, but have Mexican tolerances with respect to use on some produce without comparable U.S. tolerances;
- seventeen pesticides have tolerances in Mexico, but not in the United States (although only six of those have tolerances for food commodities that are exported from Mexico to the United States); and

- three pesticides have tolerances in both countries for the same commodities, but the tolerances are set at different levels.

Many of these tolerance differences are due to differences in weather and climate between Mexico and the United States. For these same reasons, some differences between Mexican and U.S. pesticide tolerances will always remain.

6. Industrial Chemicals

In general, Mexico follows international guidelines in determining whether to allow the use of industrial chemicals. Mexico imports most of its industrial chemicals and relies on information from the country of origin and from international organizations, including the United Nations International Registry for Potentially Toxic Chemicals, regarding the health, safety, and possible associated environmental problems of imported chemicals.

Mexico currently publishes official lists of chemicals that must be controlled when used as raw materials because of hazardous properties. Mexico also maintains a list of hazardous chemicals which are banned for use. Both of these lists accord with similar international lists.

Mexico was the first signatory to the Montreal Protocol on Substances that Deplete the Ozone Layer. SEDESOL has an especially aggressive program for reduction of chlorofluorocarbons ("CFCs") that is designed to meet or exceed the U.S. rate of reduction.

7. Divergent Standards and the NAFTA

As more is learned about environmental standards in Mexico and Canada, other significant differences with U.S. standards may be identified. Especially where divergent standards could result in significant transboundary effects on a U.S. population, the United States will work promptly to resolve such differences. Both the NAFTA itself and the Environmental Agreement provide important new mechanisms which will allow the United States to work expeditiously with Mexico and Canada to achieve higher levels of environmental protection by working toward increased "upward" harmonization of environmental standards. These mechanisms are described in detail in Sections II.A and II.B.

B. ENFORCEMENT OF MEXICAN POLLUTION CONTROL LAWS

Mexico's creation in 1992 of a new semi-independent office for environmental enforcement, the Federal Attorney General for Environmental Protection ("PFPA"), marked a significant change in the development of its environmental enforcement program. This new office has implemented a highly professional and vigorous program of inspections, leading to increasingly tough enforcement follow-up when violations are discovered. Recognizing that Mexico's General Ecology Law was only enacted in 1988, and that additional technical sophistication in the inspection program will help bridge the gap between establishment of a strong enforcement presence and widespread development of compliance-oriented environmental management practices by industry, Mexico has made impressive strides in implementing its enforcement program in just five years.

1. Background

Concerns regarding a country's potential to be a "pollution haven" have two components: the stringency of the pollution control regulations, and whether those laws are being enforced. The upward harmonization of environmental standards will not produce full environmental benefits unless the entities regulated by those standards are in compliance.

Ensuring effective environmental enforcement was one of President Clinton's principal objectives in negotiating the Environmental Agreement. The Agreement's obligations regarding effective enforcement will help to guarantee Mexico's continued progress in the development of its environmental enforcement program, while also requiring the United States and Canada to maintain and improve enforcement of their respective environmental laws. In addition, by establishing a continent-wide basis for enforcement cooperation and information-sharing, the agreement will stimulate continued and increased cooperation between the United States, Mexico and Canada in environmental enforcement.

Even without NAFTA and the Environmental Agreement, Mexico's environmental enforcement program has improved markedly in recent years. There is a strong indication that Mexico's federal environmental enforcement program, combined with increasing state and local enforcement activities, has received the attention of Mexican industry. For example, a recent survey of U.S. industries operating in Mexico, conducted by the American Chamber of Commerce of Mexico, indicates that industry holds a strong perception that enforcement measures are being more stringently applied by all levels of government than they were five years ago; and that numbers of inspections carried out by all levels of government have been increasing significantly since 1990. Additionally, inspectors have been demonstrating increasing levels of effectiveness, thoroughness, technical proficiency, and understanding of environmental laws. Evidence of recent increases in compliance with the requirement to export hazardous waste from the maquiladora industry appears to corroborate the impression that Mexico's enforcement program is increasingly effective in promoting industry compliance.

Maintaining this trend will depend upon the ability of SEDESOL, Mexico's environmental ministry, to command adequate resources to ensure that it can retain and provide continuous training for its inspectorate, and to show that it can obtain the technical sophistication to verify compliance with specific discharge and emission standards or permit conditions. The recent announcement of a $1.8 billion dollar World Bank loan to Mexico will help to ensure that these resource needs can be met in the near term.

Because information about Mexican laws and programs is not always readily available in this country, this Report presents an overview of Mexico's current environmental enforcement program, focusing on pollution control laws.[3] In addition, recent environmental enforcement activities along the U.S.-Mexico border area are outlined in Section IV.A of this report.

2. Enforcement Policies and Practices in Mexico

Any examination of the Mexican enforcement system must consider the fact that Mexico and the United States have fundamentally different legal systems and frameworks. The United States has a common law tradition, built upon reliance upon an independent judiciary to interpret law and resolve disputes among adversaries. In the United States, litigation in court plays a significant role in enforcement.

Mexico has a civil law tradition, which depends largely on administrative mechanisms and negotiation between parties to both settle disputes and enforce the law. Consequently, Mexico vests greater relative power in the executive governmental bodies to take unilateral actions, and tends to use administrative rather than judicial authority to achieve enforcement.

a. Structure of Mexico's enforcement program

Mexico's 1988 General Ecology Law vests authority to enforce environmental laws, regulations and standards primarily in the Secretariat of Social Development ("SEDESOL"). Prior to its reorganization, the top management and staff of SEDESOL's predecessor, the Secretariat of Urban Development and Ecology ("SEDUE"), had demonstrated their determination to mount a credible and effective environmental enforcement program, despite inadequate funding. This was reflected by the closure of a large number of industrial plants and facilities, including the permanent closure of a large PEMEX facility near Mexico City for failure to comply with environmental regulations and standards.

In recent years Mexico has made significant strides in enforcing its still evolving environmental regime: an increasing percentage of the environmental budget is designated for enforcement and enhancement of inspection capabilities; and the number of inspections conducted in Mexico has been increasing steadily since 1982. In order to further enhance its

---

[3] Most of this information was gathered by the Environmental Protection Agency in Mexico, through interviews with and documents provided by Mexican enforcement officials. For a comparison of Mexican environmental enforcement with U.S. practices, see the 1992 Environmental Review, which contained a detailed description of U.S. environmental laws and enforcement policies. Also, the Environmental Protection Agency is completing a report on Mexican environmental laws that will provide greater detail on Mexico's environmental enforcement program.

enforcement of environmental laws, Mexico reorganized its environmental authorities in June 1992. Mexico created within SEDESOL a semi-autonomous enforcement infrastructure, the Procuraduria Federal de Proteccion al Ambiente (the "PFPA," which roughly translates as the Federal Attorney General for Environmental Protection).

b. Inspection and enforcement procedures

Environmental enforcement in Mexico generally involves one or more of four mechanisms:

- plant closings, which may be permanent or temporary, and total or partial;
- the negotiation of compliance agreements, particularly in response to a temporary plant closing;
- the posting of a surety bond to secure compliance with an agreed or ordered schedule of compliance; and
- the imposition of fines.

These enforcement tools are implemented administratively, with SEDESOL acting both as prosecutor and adjudicator. Judicial proceedings, which would require referral of the matter to the Attorney General's Office in the Ministry of Justice, are reserved for criminal prosecutions. Although criminal actions have been rare, they have been used in response to a few incidents.

When SEDESOL investigates a facility and takes enforcement action, the investigators must strictly observe all the formalities of Mexican law. SEDESOL inspectors must document inspections by obtaining an inspection order identifying the place to be visited, the reasons for the inspection, and the scope of the inspection. The inspection order must be presented to company personnel, along with the inspector's credentials. The company must provide access to all operations and documents necessary to carry out the inspection, as outlined in the order. Refusal of access may result in the inspector requesting police assistance. Upon concluding the inspection, the inspector must prepare an inspection report on the premises.

If the inspector finds a condition or irregularity which may affect human health or the environment, the facility is notified of the initiation of administrative proceedings, and given ten days to prepare a response to the inspectors' findings. This triggers an administrative adjudication process which is generally conducted by the creation of a written record of each party's offering of proof. Through the offerings of proof, SEDESOL describes in detail the irregularities it found; the facility is given an opportunity to present rebuttal evidence; a determination is made on which counts to proceed; and needed corrective actions are identified. Within 30 days of the offering of proof, SEDESOL renders its resolution of the matter, including corrective actions to be implemented, time periods for implementation, and sanctions or penalties.

In order to levy a fine, SEDESOL must transmit its decision to the Treasury Department. If a facility shutdown is ordered, SEDESOL's decision will specify what actions must be taken before the facility may reopen, as well as any compliance or corrective requirements that must continue after reopening. Within five days of the term specified in the decision, the company must report on the status of its compliance with the decision. If a follow-up inspection uncovers non-compliance, SEDESOL may double the monetary penalty and shut down the facility, or modify a pending shut-down order to impose more stringent conditions for reopening.

Companies may petition SEDESOL for reconsideration of its decision and SEDESOL's final decision, which may sustain, overturn, or modify the original decision, must be made within 15 days of the appeal. SEDESOL's final decisions are subject to judicial review on constitutional grounds; however, because the Mexican civil law system does not rely on precedent, the courts generally grant considerable discretion to the substantive decisions of the administrative agencies, narrowly confining the scope of review to procedural irregularities which may violate constitutional protections of individual liberties.

c. SEDESOL's inspection program

SEDESOL conducts two types of inspections: (1) a multimedia, comprehensive inspection, examining the facility's total compliance with all relevant regulations and technical norms; and (2) a "short inspection", which is geared more toward determining whether paperwork requirements have been met, e.g., required facility operating licenses, annual air emissions evaluations, and monthly hazardous waste tracking records. SEDESOL's policy is to assess a fine of NP 7,000 (approx. $2,333) for each major document that is not in proper order.

SEDESOL's PFPA has initiated a vigorous inspection program, which is organized into four, essentially separate, subprograms:

1. **Targeted inspections.** Under this program, industries that pollute the most or consume large quantities of fuels are targeted for inspection visits. SEDESOL's inspectorate is instructed by PFPA headquarters in Mexico City on which types of plants present the highest priority for inspections. Top targeting priorities are companies which deal with petroleum, petrochemicals, and other hazardous substances, as well as recycling facilities. In addition, public complaints are utilized as a primary targeting device for inspection follow-up. A facility may also be targeted for comprehensive multimedia inspection based on the results of short inspections which yield suspect information. In the Federal District of Mexico, helicopter overflights provide a tool for targeting facilities, particularly for suspected violations of air quality requirements. Moreover, SEDESOL hopes that the results of its voluntary audit program, described in more detail in part (f), below, will assist in targeting inspections.

2. **Public complaints.** The establishment of a Subprocurador for Social Participation and Complaints ensures that SEDESOL is equipped to routinely investigate citizen complaints about polluting industries. Although there is no formal citizen suit mechanism comparable to that available under U.S. law, SEDESOL has made it clear that it is receptive to citizen complaints and tips

about environmental violations, and that it will respond to these complaints with plant inspections.

For example, from August 1992 to February 1993, SEDESOL received complaints about 187 facilities in the Federal District of Mexico, and responded to all of them. Of the 187 plants inspected, SEDESOL ordered total temporary shutdowns of 11 facilities, partial temporary shutdowns of 80 facilities, and made recommendations for operational or management improvements at the remaining 96 plants. SEDESOL officials report that they have received much public support for this program and hope that public participation will continue to grow as the public sees the results of the program.

3. **Aerial surveillance.** SEDESOL has instituted an aerial helicopter surveillance program in the metropolitan Mexico City area. This program takes place only in the winter, when air quality is at its lowest in Mexico City. In January to February of 1992, the helicopter surveillance program identified 148 air-polluting facilities for inspection visits. Of the 148 plants visited, SEDESOL ordered two total shutdowns, and 54 partial shutdowns. In the remaining 92 plants, SEDESOL inspectors recommended operational or management improvements to ensure compliance.

4. **Vehicle emissions.** SEDESOL, in coordination with authorities in the State of Mexico and the Mexico City Federal District government, has instituted a program to identify vehicles producing excessive air pollution. As of March 1993, 17,000 vehicles had been stopped pursuant to this program. Operators of the vehicles are fined 24 times the daily minimum wage salary in Mexico City, and given 24 hours to repair the vehicle, notified by a large sticker placed on the car. If the vehicle is stopped again within 30 days, the authorities may seize the vehicle.

In addition, SEDESOL maintains a verification program which consists of follow-up on previously-visited facilities.

Shortly after its creation, SEDESOL developed a work plan for conducting inspections in the Federal District of Mexico, calling for 200 inspections per month. This work plan was initiated in August 1992. SEDESOL was able to exceed its goal of 200 inspections per month, and in November 1992, modified the work plan to set a goal of 500 inspections per month. This goal was immediately exceeded: in December 1992, SEDESOL conducted 1000 inspections in metropolitan Mexico City, while in January 1993, 1008 such inspections were conducted. In the year from August 1992 through August 1993, SEDESOL conducted 8,304 inspections in the city, exceeding its target by more than 3,000 inspections. These resulted in temporary partial closures of 522 facilities and temporary total closure of 29 facilities.

Having successfully implemented the inspection work plan in greater Mexico City, SEDESOL extended the work plan to the rest of the country, calling for 750 inspections per month in areas outside of greater Mexico City. From its creation in June 1992 through September 1993, SEDESOL's PFPA has carried out approximately 16,386 inspections, resulting in 1161 temporary partial closures, 216 temporary total closures, and over 100 permanent plant shut-downs. Two thousand four hundred and forty-seven (2,447) of these inspections were carried out on the Mexican side of the U.S.-Mexico border area, resulting in 202 temporary partial closures and 58 temporary total closures.

One explanation for SEDESOL's early success in meeting its aggressive inspection goals is that many of the inspections were of the "short inspection," paperwork-violation variety. However, SEDESOL's focus on paperwork requirements is in response to allegations that its predecessor agency, SEDUE, lacked basic information about the number of facilities operating subject to its regulatory jurisdiction. SEDESOL has initiated an entirely new program, geared first toward achieving compliance with facility authorization and other information requirements which form the fundamental bases of pollution control. In so doing, SEDESOL has set in process a much-needed effort to establish baseline data on industrial operations nationwide, while promoting compliance among industrial facilities in Mexico.

SEDESOL inspectors also conduct comprehensive, multimedia inspections. Currently, these inspections are largely visual, and do not involve actual sampling of emissions or discharges. Rather, inspectors examine inventories of chemicals used and released, or inspect to determine whether emission control technologies have been adopted. At times, actions may be taken based on visual observations of actual pollutants, such as for opacity violations or discharges with obviously noxious odors.

SEDESOL reports that as of October 28, 1993, it has 90 inspectors in the Mexico City metropolitan area, 130 inspectors in the U.S.-Mexico border area, and a total of 460 inspectors country-wide. (Forty percent of Mexico's industrial operations are located in the Mexico City metropolitan area.) These figures reflect that SEDESOL's inspectorate, particularly in the border area, has experienced some attrition, which has been attributed to SEDESOL's inability to compete with private sector salaries, as firms gear up to hire environmental managers. SEDESOL is working to address this problem by developing a permanent training program to maintain a sufficient number of trained inspectors, with its new world bank funds, it hopes to increase the total number of inspectors country-wide to 600 within the next four years.

SEDESOL hopes to use funds from its World Bank loan to obtain equipment for sampling and analyzing pollutants. Such technology is much-needed to enable SEDESOL inspectors to move fully from the first phase, that of establishing compliance with baseline authorization and recordkeeping requirements, to a full compliance inspection program capable of detecting violations of numerical protection standards.

In implementing many of its environmental requirements, SEDESOL relies on shifting the burden to the regulated facility to analyze and document releases to the environment, or to install specific pollution control equipment. SEDESOL can thus base enforcement responses on whether facilities have installed the required equipment, or upon the completeness and integrity of monitoring or materials usage data. For example, although SEDESOL has not yet published emission standards for volatile organic compounds ("VOCs"), it is requiring companies to measure VOC emissions.

These practices are somewhat comparable to those in the United States. For example, enforcement of the U.S. Clean Water Act relies heavily on discharge monitoring reports submitted monthly by facilities. Actual discharge sampling is uncommon except to verify violations discovered through the review of such reports. Under the Clean Air Act, actual stack testing is rare except when a facility first installs equipment. Inspections are geared to verifying compliance with requirements for operating and maintaining such equipment. The 1990 amendments to the Clean Air Act place increasing emphasis on self-monitoring requirements. Under the Resource Conservation and Recovery Act, inspections rely on visual verifications of facility management practices and review of records, including groundwater data from samples collected and analyzed by the facility, with actual inspector-conducted sampling limited to investigations into actual releases of hazardous waste constituents to the environment.

Mexico also uses negotiated compliance agreements to implement its environmental requirements. These agreements may include provisions requiring companies to install VOC recapture equipment, or other emission control devices. Similarly, a number of companies in metropolitan Mexico City recently committed to switching to natural gas fuel to reduce emissions. Although this was a voluntary action, the companies have signed an agreement with local government authorities committing to the change, thereby creating an enforceable voluntary compliance agreement. In addition, although the use of natural gas in Mexico City is voluntary for existing facilities, it is mandatory for new industries.

d. Sanctions and settlement negotiations

A primary enforcement tool utilized by SEDESOL is the plant closure, which may be temporary or permanent, and may involve closing the entire facility or only a portion of its operations. A closure order results when SEDESOL inspectors discover a direct and significant threat to the environment or human health, or a high level of nuisance, such as noise pollution.

Temporary closures are ordered when the immediate problem creating a health or environmental threat is remediable. Such temporary closings are intended to lead to consultations between SEDESOL and the corporate entity involved. The closings occur in advance of negotiations, and the plant is allowed to reopen only after the company resolves the immediate problem, and an agreement with timetables for achieving full compliance is reached. In negotiating a compliance agreement, SEDESOL may use its discretion to allow industries, especially smaller industries, a reasonable time to comply with its requirements. These legally enforceable agreements are monitored by SEDESOL.

Previously, SEDESOL relied heavily on requirements that facilities post performance bonds to ensure their compliance with the terms of a settlement or order. Once the performance bond was created, SEDESOL would allow a closed plant to reopen. SEDESOL discovered, however, that the bonds often did not provide facilities with adequate incentive to stick to the compliance agreement. Hence, as of August 1992, SEDESOL has tightened its policy to require the actual correction of problems prior to lifting a shutdown order.

Permanent closures are employed more rarely than temporary closures, but the threat of permanent closure serves as a major deterrent to noncompliance. A permanent closure might be ordered where a facility has very significant emissions problems, perhaps exacerbated by the fact that it is located in a highly-populated area where exposure risks are increased. On this basis, Mexico in 1991 permanently shut down a major facility of PEMEX (the state owned oil company). Permanent closures are likely to be ordered when the problems are impossible or too expensive to fix; however, this sanction may also be imposed punitively, such as for a history of extreme noncompliance, even where mitigation of the immediate environmental risks created by the plants' operations might be possible.

SEDESOL prefers, however, to order total, but temporary closure, and negotiate solutions wherever possible that will allow the plant to reopen. At times, plants subject to temporary closure orders may terminate operations permanently if the operational changes sought by SEDESOL are too expensive to implement. When a plant closes permanently and relocates, it will be subject to all SEDESOL requirements for new operations, including permit requirements, environmental impact assessments, and compliance with regulations and ecological norms.

Both SEDESOL and the U.S. Environmental Protection Agency rely on negotiated settlements to achieve compliance and remediation of environmentally unsound conditions. Approximately 95 percent of EPA's administrative and civil judicial actions are concluded as negotiated settlements. The primary distinction between the Mexican system and the U.S. system is that in Mexico, a strong sanction (facility shutdown) is imposed prior to initiation of negotiations, and continues in effect until negotiations are complete and the facility undertakes agreed-to corrective measures to SEDESOL's satisfaction.

In the United States, for a court or administrative tribunal to enjoin continuing activities of the subject of an enforcement action prior to full adjudication or settlement of the matter, the government must satisfy a high burden of demonstrating that the company's actions present an imminent and substantial endangerment to human health and the environment, or that emergency injunctive relief is otherwise warranted. In the Mexican system, the authority to shut down a facility pending the negotiation of a compliance agreement provides SEDESOL with substantial bargaining power in promoting rapid and favorable settlements.

Current Mexican law permits SEDESOL to impose fines equivalent to between 20 and 20,000 times the daily minimum wage in the Federal District of Mexico (up to approximately $85,000, as of January 1993). Fines can be imposed on a per-day, per-violation basis for as long as the violation persists. By comparison, most U.S. environmental statutes allow for civil penalties of up to $25,000 per day per violation. Thus, the range of monetary fines that can be imposed by SEDESOL is comparable to, and can even exceed, those imposed by the United States.

In practice, just as in U.S. administrative and civil judicial enforcement cases, the penalties sought or agreed to in settlement for initial violations may be much lower than the statutory maximum, based upon considerations of economic fairness and the seriousness of the violation. However, SEDESOL may double the fine for persistent violations, creating a powerful deterrent against failing to implement the terms of an agreed or ordered compliance schedule.

Another Mexican environmental enforcement tool is administrative detention, which can result in the deprivation of a corporate officer's freedom for up to 36 hours. More commonly, it is applied for several hours on a daily basis until agreement is reached on a compliance plan and schedule.

Certain environmental regulations contemplate criminal prosecutions. SEDESOL may refer a criminal case to the Federal Attorney General to initiate prosecution at any time it believes evidence of a crime exists, including during the conduct of administrative enforcement proceedings. Such proceedings, although rare, have been increasing, particularly in cases involving hazardous waste disposal, where disposed wastes may provide clear evidence of patently criminal activity. In a recent case, for example, the operator of a solvent recycling facility was charged with mismanagement and illegal disposal of wastes brought onsite despite inadequate recycling capacity. The operator was arrested, placed into custody, and released on a $1 million bond which secured his cooperation in cleaning up the waste.

e. SEDESOL oversight of other federal and state agency enforcement

SEDESOL shares jurisdiction over water quality protection with the National Water Commission ("CNA"). While CNA has jurisdiction over pollution of national waters, SEDESOL may promulgate technical norms pertaining to the discharge of hazardous or toxic wastes into water. CNA has primary authority to enforce the standards promulgated by SEDESOL, and can itself create special conditions on these discharges as long as they are no less stringent than SEDESOL's requirements.

SEDESOL, however, plays an important function in overseeing and monitoring CNA's enforcement of water quality regulations and norms. SEDESOL inspectors are trained to evaluate water discharges for compliance, and may receive complaints from the public about industrial discharges to water. SEDESOL refers violations detected by its inspectors or alleged by citizen complaints to CNA for follow-up. If the CNA does not respond to these claims, SEDESOL's PFPA can counsel CNA about surveillance of water discharges, and make recommendations. If this does not result in satisfactory action by CNA, SEDESOL may bring a claim to the General Comptroller's office alleging that CNA has failed to discharge its duty. Such claims are litigated in accordance with federal law regulating the responsibility of public servants, with sanctions including fines, removal of officials, and possible criminal action.

SEDESOL's role as monitor of CNA's enforcement of SEDESOL water norms applies to other federal agencies as well. The PFPA often receives claims that concern the work of other federal agencies, and has similar authority to ensure that these agencies comply with SEDESOL requirements. SEDESOL reports that its recommendations are generally followed, and that it has not been necessary to resort to formal legal proceedings before the Comptroller General for SEDESOL to ensure that other agencies carry out its mandate.

In addition, SEDESOL officials report that, on an operational level, they are working cooperatively with CNA to ensure joint compliance monitoring of industry. The two agencies have conducted several inspections jointly.

SEDESOL, at least theoretically, enjoys a similar oversight role with respect to state implementation of its environmental laws. Many states have begun to promulgate their own environmental laws. SEDESOL has the authority to verify state enforcement of its environmental laws, and may make recommendations to ensure that state laws implementing federal standards are adequately enforced. However, in practice, while many states have passed environmental laws, some of these states have discovered that they are not prepared to fully implement these laws. The result is that, at the moment, the federal government is more involved in direct enforcement than in overseeing state enforcement activities.

f. SEDESOL's environmental audit program

In addition to its program of targeted inspections and responses to public complaints, SEDESOL has initiated an innovative program of voluntary environmental audits, promoting compliance by allowing facilities an opportunity to discover irregularities in their operations, and eliminate them prior to the appearance of an inspector and the threat of shutdown or fines. SEDESOL developed its program, and built its capacity to implement the program, by consulting with and participating in training exercises provided by a number of governments and private institutes in North America and Europe.

The program is targeted toward high-risk industries, and designed to identify risks created by operations likely to lead to environmental accidents or contamination, and potential compliance irregularities. Accordingly, the audit serves a dual function of encouraging individual facility compliance, as well as providing SEDESOL with a baseline of background information on management practices in high-risk industries for the future targeting of priorities for its inspection program.

Environmental audits are conducted by approved private consultants. The audit entails a comprehensive plant survey, conducted in three phases:

1. **Pre-Audit.** The auditor reviews basic data, including facility hazardous materials handling records, as well as worker safety and health records, and develops a work plan for conducting the audit.

2. **Audit.** The audit itself thoroughly evaluates internal management of the plant, including company environmental policy, hazardous waste handling practices, emergency response mechanisms, and other factors pertaining to controlling pollution or contamination.

3. **Post-Audit.** The audit results are evaluated, and based on the results, an action plan is developed for correcting each of the problems identified. SEDESOL then negotiates the action plan with the facility, which then becomes legally binding and may be secured by a performance bond.

The action plan becomes, in essence, an enforceable contract between the facility and SEDESOL. If the facility fails to comply fully with the action plan, SEDESOL may declare the performance bond forfeited, and also may institute formal inspection and administrative proceedings to penalize or shut down the facility for persistent violations discovered by the audit and not corrected. In addition, if, during the conduct of the audit, the auditor discovers serious violations of the General Ecology Law or its regulations and norms, SEDESOL may request immediate corrective action or cessation of the problematic activity -- and may, upon following procedural requirements for inspections and administrative enforcement, impose the full range of its enforcement sanctions if its request is not followed.

Although the audit may lead to enforcement measures, a company's decision to participate in the audit program is purely voluntary. As a result, SEDESOL is not required to obtain an inspection order to conduct an audit, and the audit itself will not lead directly to the imposition of a penalty or shut-down order. Administrative enforcement procedures would have to be initiated and followed in the event uncorrected violations discovered by the audit require the imposition of enforcement sanctions.

SEDESOL initiated the audit program in late 1992, targeting for participation petroleum extraction industries, as well as petroleum product, petrochemical, and textile manufacturers. The initial effort focused on the State of Veracruz, with 19 audits conducted in the petroleum and petrochemical industry concentrated there. Action plans have been finalized with at least four of these companies, and are being negotiated with the remaining companies. Fifty-eight more audits were then conducted throughout the country, covering maquiladora facilities in border states, as well as companies in the Valley of Mexico, Guadalajara, Monterrey, and Michoacan.

SEDESOL financed the initial 77 audits, but hopes to gradually phase out its financing as the program attracts voluntary participation by companies that will fund and conduct their own audits using approved independent consultants. After approaching the initial companies to encourage participation in the audit program, SEDESOL has embarked on the next phase of the program, in which it will assess the degree to which companies will participate without being directly solicited. Because the program focuses on companies in high-risk industries or zones, which are targeted for regular inspection and the subject of common public complaints, SEDESOL believes that the threat of inspection and its possible sanctions will serve as an incentive for voluntary participation. In addition, SEDESOL expects public corporations to embrace environmental audits: PEMEX, for example, has agreed to audit all of its facilities.

Beyond their voluntary nature, the audits differ from inspections in that they are more flexible: in addition to identifying current violations, audits also attempt to identify areas which may lead to potential violations or risks to human health and the environment. Thus, the audit program is comparable to a preventive check-up.

The audit program also has the potential to promote the adoption of practices or operational changes which go beyond mere compliance in reducing pollution emissions (i.e., classic "pollution prevention" as defined in the United States). However, this potential is currently limited by two factors. First, because of its voluntary nature, facilities are only likely to adopt process or practice changes which go beyond mere compliance if the auditor and/or SEDESOL is able to convince them that such changes are within their economic self-interest. Thus, facilities are most likely to agree to process changes recommended by the auditor only when they would also increase efficiency. Second, the audit program's newness and voluntary nature may discourage SEDESOL from assertively negotiating innovative action plans. For the present, it may be more inclined to focus on the basic goals of achieving compliance and preventing extraordinarily risky practices leading to risks of accidental exposure to contaminants, rather than seeking radical changes to plant operations.

Despite these constraints, SEDESOL's audit program is clearly innovative, going beyond the traditional exercise of enforcement functions as utilized in the United States by actively promoting compliance and sound environmental management practices through preventive, non-adversarial facility evaluations. At the same time, the audit program will serve as a primary source of information on which to base future inspection targeting. As an ancillary benefit, once fully accepted by industry, the audit program has great potential to lead to the identification, and potential negotiation of binding agreements for, changes to facility processes and practices which will reduce pollution beyond the levels required to achieve mere compliance and minimize the risks of extraordinary accidents.

IV. RECENT DEVELOPMENTS IN THE U.S.-MEXICO ENVIRONMENTAL RELATIONSHIP

The United States and Mexico have a long history of cooperation on environmental issues. These joint efforts were given added impetus by the NAFTA negotiations and the negotiation of the related environmental agreements. Implementation of the NAFTA can be expected to further strengthen the commitment of both the U.S. and Mexican governments to these programs.

Mexico's commitment to environmental protection is further demonstrated by the number of international environmental and conservation agreements to which it is a party (see Table 1 on page 48).

The 1992 Environmental Review outlined the history of U.S.-Mexican environmental cooperation, and described a number of new joint environmental initiatives. The following sections describe recent developments in these programs.

A. INTEGRATED BORDER ENVIRONMENTAL PLAN

The Integrated Border Environmental Plan ("Border Plan"), which was issued in February 1992, is intended to provide for the long-term protection of human health and the environment within the U.S.-Mexico border area. The Plan contemplates a multi-phase process to achieve this goal. The Border Plan has the following four objectives:

(1) to strengthen enforcement of existing environmental protection laws;

(2) to reduce pollution and improve the quality of the border area through new initiatives;

(3) to increase cooperative planning, training and education; and,

(4) to improve understanding of the border area environment.

The two agencies responsible for coordinating activities under the Border Plan are the U.S. Environmental Protection Agency ("EPA") and Mexico's Secretariat of Social Development ("SEDESOL"). However, a number of other agencies in both countries are participating in the Border Plan, including Mexico's National Water Commission, Civil Protection of Mexico, the Mexico and U.S. Sections of the International Boundary and Water Commission ("IBWC"), the U.S. Customs Service, and various U.S. and Mexican state agencies.

In the 20 months since it was announced, Mexico and the United States have been successful in achieving many of the objectives of the Border Plan. Following is a summary of progress to date on the implementation of the Border Plan. The determination of both countries to respond to public concerns in the border area is demonstrated by the fact that much has been accomplished with limited resources.

TABLE 1

MEXICAN ADHERENCE TO MULTILATERAL ENVIRONMENTAL AND CONSERVATION AGREEMENTS

In addition to having adhered to politically-binding documents such as the Rio Declaration and Agenda 21, Mexico is also a party to numerous multilateral environmental and conservation agreements, including:

- the Convention for the Protection of the Ozone Layer (the "Vienna Convention");
- the Montreal Protocol on Substances that Deplete the Ozone Layer (the "Montreal Protocol");
- the Convention on Nature Protection and Wildlife Preservation in the Western Hemisphere (the "Western Hemisphere Convention");
- the Convention on International Trade in Endangered Species of Wild Fauna and Flora ("CITES");
- the Convention on Wetlands of International Importance, Especially Waterfowl Habitat (the "Ramsar" Convention);
- the Convention on Fishing and Conservation of Living Resources of the High Seas;
- the International Convention for the Prevention of Pollution of the Sea by Oil;
- the International Convention Relating to Intervention on the High Seas in Cases of Oil Pollution Casualties;
- the Convention on the Prevention of Marine Pollution by Dumping of Wastes and Other Matter (the "London Convention");
- the Protocol relating to Intervention on the High Seas in Cases of Pollution by Substances Other than Oil;
- the Protocol of 1978 relating to the International Convention for the Prevention of Pollution from Ships;
- the Convention for the Protection and Development of the Marine Environment of the Wider Caribbean Region (the "Cartagena Convention");
- the Protocol Concerning Cooperation in Combating Oil Spills in the Wider Caribbean Region;
- the International Convention for the Regulation of Whaling; and
- the Convention Concerning the Protection of the World Cultural and Natural Heritage (the "World Heritage Convention").

Further, Mexico has ratified the United Nations Framework Convention on Climate Change, the convention on Biological Diversity, and the United Nations Convention on the Law of the Sea. (These agreements are not yet in force.)

Early in 1994, EPA and SEDESOL will begin developing a Border Action Program that will take the place of the current Border Plan. The Program will encompass the activities of other U.S. agencies with important environmental and conservation programs in the border area, including the Department of the Interior and the Department of Health and Human Services.

1. Strengthen Enforcement of Existing Laws

The effort to strengthen enforcement of existing U.S. and Mexican laws has focused on increasing inspection activities on both sides of the border, and improving coordination between U.S. and Mexican enforcement agencies. Specific areas targeted by this effort are outlined below.

- Expand enforcement efforts:

  SEDESOL has launched a comprehensive inspection program. Since June 1992, over 16,000 inspections have been carried out, including 2,447 in the border area. These inspections resulted in temporary partial closure of 1,161 companies (of which 202 were located in the border area), and 216 temporary total closures (of which 58 were in the border area).

  In the U.S. border region, preliminary estimates for FY 1992 and 1993 indicate that more than 351 EPA inspections and 1,671 state inspections were carried out, and 15 civil judicial enforcement actions, 114 administrative actions, and three EPA-led criminal prosecutions were initiated under the federal air, water, and hazardous waste statutes.

  In June 1992, Mexico and the United States initiated numerous enforcement actions targeted at the border area. Mexico conducted 42 inspections, discovering violations at 22 facilities which resulted in eight shut-down orders and four bond forfeitures. The United States undertook 17 enforcement actions, seeking $2 million in penalties and two criminal indictments. The United States also filed four enforcement cases under the Resource Conservation and Recovery Act ("RCRA") which were developed using the binational Hazardous Waste Tracking System pilot data base.

  In the United States, transboundary surveillance efforts have been intensified. A task force of U.S. federal and state agencies in California has increased detection of illegal transboundary shipments of hazardous wastes. California and Texas have provided personnel to conduct routine inspections of transboundary shipments at border crossings. EPA is providing funding to state environmental agencies to expand this capability for routine surveillance and detection of transboundary shipment violations along the border.

- Establish Cooperative Enforcement Strategy Working Group:

  Through the work of this group, Mexico has assisted the United States in developing a group of related cases enforcing U.S. environmental import and export laws. Eight of these cases involved violations regarding hazardous waste exports to Mexico.

In addition, EPA and SEDESOL have established Regional subgroups to cooperate in border enforcement activities, providing a forum for exchanging information on alleged violations and targeting common environmental problems for enforcement response.

- Cooperate on enforcement actions in priority areas:

  Case-specific cooperation between the United States and Mexico has resulted in the initiation of clean-up at illegal waste sites in Mexico and/or investigations into potential violations of U.S. or Mexican law in at least six separate instances.

  U.S. and Mexican Customs officials have begun coordinating inspection and safety techniques. Binational customs coordination has resulted in two criminal indictments for illegal waste exports, and halted several attempted illegal waste shipments.

  In addition, a cooperative pilot effort has been undertaken to encourage voluntary compliance with Mexican laws by U.S. firms with subsidiary operations in Mexico. Several U.S. companies have pledged to ensure that their Mexican subsidiaries comply with Mexican law. Some companies have submitted data evidencing a compliance review. More recently, SEDESOL has implemented an innovative voluntary environmental audit program (described in more detail in Section III.B).

- Build enforcement capacity:

  EPA conducted six "Multi-Media" Inspector Training courses, providing training in inspection techniques to 370 SEDESOL inspectors (including 200 in the border area) from March to September 1992. SEDESOL inspectors have also received media-specific inspection training. Two SEDESOL students received training in the use of aerial surveillance to identify potential waste disposal sites. A number of SEDESOL inspectors have attended EPA-sponsored training in hazardous waste and air pollution inspection techniques.

  EPA has provided transboundary waste shipment inspection field training for approximately 500 customs inspectors, including Mexican customs inspectors.

  More inspector training is planned, as well as delivery of EPA's Principles of Environmental Enforcement Course for SEDESOL policy-makers, and "train-the-trainer" courses to allow SEDSOL to reproduce these courses in-house for new personnel. A bilingual Customs video course for detecting illegal hazardous waste shipments is being produced, and will be delivered to U.S. and Mexican Customs inspectors in the near future.

- Increase the number of Mexican border environmental inspectors:

  SEDESOL has put in place a force of 130 environmental inspectors in the border area.

- Exchange enforcement information:

  SEDESOL has provided EPA with a summary of enforcement activities undertaken in 1992 and through August 1993, while EPA has provided SEDESOL with reports on enforcement accomplishments for 1991-1992. EPA and SEDESOL are exchanging information about their policies for assessing penalties for environmental violations.

2. Reduce Pollution Through New Initiatives

The effort to reduce pollution has focused on programs to improve water and air quality, as well as disposal of both solid and hazardous waste. Many of these programs involve the provision of technical assistance. Specific areas targeted by this effort are outlined below.

- Increase wastewater treatment:

  The United States and Mexico have initiated design, construction and/or operation of at least 13 wastewater treatment facilities in the border area. Projects are in various stages of construction in San Diego, Tijuana, Nogales, Nuevo Laredo, Laredo and several other cities.

  In addition, EPA has conducted seminars on low cost, low maintenance residential wastewater treatment systems and held workshops providing information on financial assistance to residents.

- Strengthen industrial wastewater pre-treatment programs:

  A pretreatment program has been initiated in Ciudad Juarez. The United States and Mexico have agreed in principle on a binational pretreatment program for the Nogales International Wastewater Treatment Plant. The IBWC has carried out an initial influent characterization.

- Protect drinking water supplies:

  Both governments have worked cooperatively to monitor water quality in the border area. The United States has undertaken an inventory of drinking water supply systems in the Lower Rio Grande Valley, while Mexico has inventoried water supply systems in the principal border cities of the Rio Grande/Rio Bravo.

  In addition, a binational field survey of toxic contamination of the Rio Grande/Rio Bravo has been conducted; a report should be completed by April 1994. The United States has initiated a wellhead protection program for El Paso and begun to collect data on the groundwater protection program for the Nogales area.

- Improve air quality:

  - Ciudad Juarez-El Paso: A long-term quality-assured air pollution monitoring network was established in Ciudad Juarez in June 1990. Extensive modeling activities are underway.

  - Mexicali-Imperial Valley: A study on PM-10 (respirable particulates) was initiated in 1992 to assess concentrations of particulates and to develop control strategies.

  - Tijuana-San Diego: A monitoring station located at the Tijuana Institute of Technology is operating with initial calibration complete. The station is currently monitoring NO_x, O_3, CO, SO_2, PM-10, lead, and wind speed/wind direction. A pilot vehicle inspection and maintenance program for fleet vehicles has been initiated.

  - Nogales: In coordination with SEDESOL and state/local officials, a study is being initiated to assess cross-border transport of PM-10 and air toxins in the Nogales area.

- Improve disposal of solid and hazardous waste:

  EPA and SEDESOL have developed a pilot tracking system to document and confirm transborder movement of hazardous waste from maquiladoras to U.S. treatment/disposal facilities, thereby enabling verification of compliance with U.S. and Mexican laws. EPA has provided database software and training to SEDESOL. EPA has also initiated four enforcement actions against companies with import violations identified through the data base.

  EPA has assisted SEDESOL in facilitating the safe shipment of abandoned drums of maquiladora waste to U.S. disposal facilities. EPA and SEDESOL have also exchanged information on maquiladora facilities that generate hazardous wastes.

- Promote pollution prevention:

  In 1992, the United States and Mexico conducted twelve joint cooperative pollution prevention training visits to facilities in Texas and Arizona.

  The Pollution Prevention Work Group has completed a Bilingual Pollution Prevention Manual for the metal finishing industry, and is beginning work on a manual for the wood finishing industry.

- Provide other technical assistance:

  After an on-site review and evaluation of the SEDESOL computer system by EPA, SEDESOL and EPA are working to overcome some compatibility problems with information exchange and access to environmental data bases.

3. Increase Cooperative Planning, Training and Education

Programs in this area have focused on emergency preparedness and response, and cooperative training efforts. Specific program areas are outlined below.

- Increase training for emergency preparedness and response:

  The United States and Mexico are cooperating to enhance emergency response capabilities through a Joint Response Team ("JRT"). A work group of state and national experts has sponsored meetings with officials from both countries. The work group has completed training for public officials in the Sister Cities on hazardous materials and crisis management.

  The United States and Mexico are also drafting protocols to facilitate cross-border mobility of emergency response equipment and personnel. Many of the Sister Cities have used informal agreements successfully along the border.

  In addition, the United States and Mexico have improved methods of making data and information accessible and available to Sister Cities concerning hazardous substances. EPA has completed Sister City profiles for Region 6, which include information on hazardous materials at fixed facilities and in transport.

  Sister City Contingency Plans have also been developed. Plans have been completed for Brownsville-Matamoros and Imperial County-Mexicali. Plans are near completion in San Diego-Tijuana, Nogales-Nogales, and Nuevo Laredo-Laredo. The United States and Mexico have held exercises of the Calexico-Mexicali plan, San Diego-Tijuana plan, Brownsville-Matamoros plan, Ciudad Juarez-El Paso plan, and Nuevo Laredo-Laredo plan.

- Promote environmental education and increase technical capabilities:

  Air quality training has been provided on such topics as ambient air monitoring, air dispersion modeling, emissions inventories, laboratory quality assurance, air quality management, air quality meteorology, and combustion evaluation.

  In addition, training has been provided on hazardous waste management, as well as landfill site selection and facility management. EPA and SEDESOL have conducted cooperative training visits to border facilities.

  Finally, as noted above, training has been provided to 370 SEDESOL environmental inspectors in the techniques of conducting multimedia compliance inspections.

4. Improve Understanding of the Border Environment

Programs in this area have focused on preparing inventories of waste disposal sites and shared water resources, as well as monitoring air quality and preparing periodic reports on the border environment.

- Inventory waste disposal sites:

  The United States and Mexico have agreed to a consultative mechanism for the exchange of information on existing and proposed hazardous waste sites in the border zone.

  In order to locate abandoned or illegal hazardous waste dump sites, EPA has developed strategies for the border area using available Geographic Information Systems ("GIS") technology and aerial surveillance, as well as information from local agencies. EPA has provided training on aerial photography interpretation, while SEDESOL is designing a strategy to address the problem of abandoned waste sites.

  Finally, EPA has compiled extensive data on municipal solid waste disposal facilities and completed an inventory of U.S. sites. Municipalities in Mexico are conducting an assessment of municipal solid waste disposal facilities along the Mexican border area.

- Inventory shared water resources:

  Field work has been completed on a binational survey of the Rio Grande/Rio Bravo for toxic pollutants, and a binational report is under development. A similar study for the Colorado River is being considered.

- Monitor air quality:

  Emissions inventories and monitoring networks are being developed for priority binational air basins to determine ambient air pollution concentrations, apportion sources and their relative impacts, recommend cost effective control strategies, and measure progress and compliance.

- Prepare periodic reports on the border environment:

  A catalog has been prepared, based on GIS data, describing the environment of the U.S. border area.

B. TRAINING AND EDUCATION

During the last year and a half, EPA, the U.S. Fish and Wildlife Service ("USFWS") and the National Oceanic and Atmospheric Administration ("NOAA") have worked with SEDESOL and other Mexican agencies on a number of cooperative training and education activities. These activities have focused on environmental enforcement, pollution monitoring and control, and management of nature reserves and marine resources. While many of these programs have been designed for Mexican environmental officials, other programs have been targeted at maquiladoras, local community officials, and the public.

Moreover, in September 1992, the heads of EPA, SEDESOL, and Environment Canada signed a trinational agreement to promote environmental education in all three countries.

Two workshops have been convened and a trilateral committee has been established to facilitate information exchange and to develop joint programs.

1. Enforcement of Environmental Laws

To improve the enforcement work force along the U.S.-Mexico border, EPA has provided training for U.S. and Mexican inspectors, as well as customs officials. These training programs are described in more detail in Section IV.A.

The U.S. Fish and Wildlife Service has presented training seminars on wildlife enforcement and on activities associated with the Convention on International Trade in Endangered Species of Wild Fauna and Flora ("CITES") to over 100 representatives of SEDESOL, as well as to other Mexican officials responsible for controlling trade in wildlife. Partly as a result of the training, several individuals were recently found guilty of smuggling ocelots, jaguars, parrots, palm cockatoos, and over 1000 bobcat hides from Mexico.

2. Air Quality Monitoring and Management

EPA training courses have covered many important topics in air monitoring, air dispersion modeling, laboratory quality assurance, air quality management, air quality meteorology, and combustion evaluation. EPA assisted SEDESOL in defining "easily implementable controls" and alternative controls. EPA's Office of Air Quality Planning and Standards has provided a technical staff person on a detail to SEDESOL to provide long-term technical support in developing emissions inventories.

NOAA's Forecast Systems Laboratory and Mexico's Centro de Ciencias de la Atmosphera are considering initiating a study that would enable improved forecasting of meteorological conditions over the Valley of Mexico and the surrounding mountains during conditions of severe pollution episodes. Such a study would enable U.S. and Mexican officials to better predict the atmospheric consequences of alternative strategies for reducing the pollution associated with economic development in the area.

3. Marine Pollution Monitoring

NOAA is examining means to improve the exchange of information concerning marine pollution monitoring programs being conducted by Mexico and the United States. In addition, NOAA is exploring opportunities for coordinated monitoring activities in the U.S.-Mexican border environment. Wider distribution of such data among Mexican government officials and academics should contribute to the improvement of pollution control and fisheries management along the U.S. and Mexican coastlines.

4. Other Pollution Control Programs

Pesticides. At the request of SEDESOL, EPA has begun to develop bilingual training and outreach programs on border pesticides-related issues for the Lower Rio Grande Valley (Texas-Tamaulipas) and Imperial Valley-Mexicali areas.

Hazardous and solid wastes. EPA training has covered the transboundary movement of hazardous waste, hazardous waste management and inspection techniques. EPA and SEDESOL have conducted cooperative training visits to border facilities. In addition, training programs have been developed and planned for landfill site selection and facility management.

Emergency preparedness and response. The United States and Mexico have worked together to identify and conduct appropriate workshops and training sessions, and to provide technical assistance to the Sister Cities along the border. These activities are described in more detail in Section IV.A.

Environmental impact assessments. Twenty SEDESOL personnel attended an EPA course on environmental impact assessment in December 1992. This intensive course, prepared specifically for an international audience and modified for Mexico, was facilitated by EPA staff using simultaneous Spanish translation. In September 1993, the U.S. Agency for International Development ("AID") agreed to provide additional funding to train SEDESOL officials as course facilitators.

5. Reserve Management

The Tijuana River National Estuarine Research Reserve ("TRNERR"), located at the coastal end of the Tijuana River, is managed by the National Ocean Service of NOAA. This 2,500 acre site is adjacent to the U.S.-Mexico boundary. NOAA has worked with Mexican officials on education and research projects to monitor water quality, channel fishes, invertebrates, vegetation, and marsh soils. The objective of these research efforts is to document pollutants that enter the TRNERR with freshwater inflows from the river and to assess changes in environmental indicators.

The USFWS, together with the U.S. Forest Service, has provided short-term training courses and workshops to officials directly responsible for managing protected areas in Nuevo Leon and Yucatan.

6. Protection of Marine Wildlife

With support from the U.S. AID, the National Marine Fisheries Service is working with the Government of Mexico and the Mexican Fisheries Ministry to provide technical assistance and training for the protection of marine turtles incidentally caught in shrimp fisheries. The use of Turtle Excluder Devices and dolphin-safe polices will assist the Mexican fishing fleet in complying with the environmental and legal instruments of both countries.

In addition, the Southwest Fisheries Science Center ("SWFSC") is undertaking a number of cooperative activities with Mexico to protect the vaquita, an endangered marine mammal. The vaquita, which is listed under the Endangered Species Act, is endemic to the northern Gulf of California and has one of the most limited distributions of any marine mammal. However, it is vulnerable to incidental mortality in commercial fisheries, such as the gill net fisheries for totoaba and shark, and in shrimp trawls. The SWFSC is currently working with Mexican scientists to conduct ship and aerial surveys of the vaquita. The results of these surveys should assist in the design and implementation of more effective preservation efforts.

7. Environmental Education Programs

EPA has provided grants to several non-governmental organizations to conduct environmental education programs. One example is Project del Rio, a cooperative water quality monitoring program involving U.S. and Mexican high school students. Under the grant, Project del Rio will develop a bilingual citizen's guide to the Rio Grande watershed. Materials developed under this program will be available for use in schools as well as for the general public.

Environmental education has been a key component of many fish and wildlife cooperative programs with Mexico. Examples of such projects include a program in the Centla marshes of Tabasco, Mexico's largest estuary; a project for school children on sustainable resource management in the state of Quintana Roo; and a workshop for teachers, resource managers, and community leaders to support Mexico's newly declared Sierra de Huautla Reserve.

C. CONSERVATION

The United States and Mexico have a long history of cooperating on wildlife protection and the conservation of natural resources. By virtue of their long common border, migration patterns, and distribution of many species, both countries have a common interest in a great number of wildlife and natural resource issues. The entire border region, and in particular the Rio Grande Valley, has a great deal of unique biological diversity which the two sides are taking steps to protect.

The history of U.S.-Mexican cooperation on conservation issues was outlined in detail in the 1992 Environmental Review. Major areas of ongoing cooperation include wildlife conservation, management of national parks and wildlife refuges, and conservation of marine resources.

1. Wildlife Conservation

The majority of cooperative fish and wildlife activities between the United States and Mexico are conducted under the auspices of the Mexico-U.S. Joint Committee for the Conservation of Wild Flora and Fauna. Currently, the joint committee meets about once a year to review both policy issues and specific conservation projects. Conservation projects in the fields of training, protected areas management, endangered species, law enforcement, and migratory birds have been implemented throughout Mexico.

a. Enforcement

The entire border area experiences legal, as well as illegal, commercial plant and wildlife traffic. In 1991, Mexico became a member nation of the Convention on International Trade in Endangered Species ("CITES"). Through CITES, both the United States and Mexico are working more closely to increase wildlife enforcement along the border to meet the Environmental Agreement objectives of enhancing environmental compliance and enforcement of environmental laws and regulations.

The USFWS has wildlife inspectors stationed at several ports of entry to inspect wildlife shipments. Within the border area, there are ten "designated" wildlife ports of entry. Mexico has plans to increase its wildlife enforcement along the border.

b. Endangered species

Within 25 miles of the border in Texas, New Mexico, Arizona, and California, the USFWS has the responsibility to maintain and to seek the recovery of at least 460 species of plants and animals that are endangered, threatened, proposed, or candidate species for listing under the Endangered Species Act. Arizona and the Mexican State of Sonora share the Sonoran Desert, harboring scores of threatened and endangered species like the Sonoran pronghorn antelope, Yaqui catfish, and Cochise pincushion cactus. Cabeza Prieta National Wildlife Refuge and its Mexican "sister", Parque del Gran Desierto del Pinicate, protect portions of this ecosystem on opposite sides of the border.

The USFWS is increasing coordination with Mexico to protect and recover endangered species. For example, in April 1992, the USFWS proposed experimental reintroduction of captive-reared Mexican wolves into suitable historical habitat; a draft environmental impact statement is being prepared to address this issue. The captive breeding program was established in the 1970s with five wild-caught Mexican wolves from Durango and Chihuahua, Mexico. Currently, the captive population, which has expanded to 70 wolves, is maintained at nine facilities in the United States and three facilities in Mexico. Reintroduction sites are currently being evaluated in Arizona and New Mexico.

To ensure that no species becomes extinct, status surveys are undertaken to identify a species' status and threats to its existence. Such surveys are done in cooperation with the Mexican government. Recent surveys have assessed the status of the Sonoran pronghorn antelope, Mexican spotted owl, desert tortoise, maroon-fronted parrot, aplomado falcon, scarlet macaw, several species of desert fishes, and several cacti and other desert plants.

Since many listed species have ranges which include both Mexico and the United States, a coordinated recovery effort is essential to the maintenance and eventual delisting of many species. NAFTA could result in increased joint cooperation to accelerate development and implementation of recovery plans on transboundary species. In one example of current joint efforts, working with SEDESOL, Mexico's Fisheries Ministry ("SEPESCA"), and Mexican universities, the USFWS has aided the recovery of several sea turtle species, including efforts at nesting beaches for the Kemp's Ridley turtle in Tamaulipas and the black turtle in Michoacan.

In another example, in 1991, Cabeza Prieta National Wildlife Refuge established a core working group which includes all the land management entities and agencies within the historic range of the Sonoran pronghorn antelope. The USFWS, the Bureau of Land Management, U.S. National Park Service, the Government of Mexico, the Tohono O'Odam Indian Nation, and the Arizona Game and Fish Department developed an action plan to implement recovery actions for this species with initial priority given to accomplishment of a range-wide survey of the species, to be initiated in 1992-93. The survey has found a new population in Sonora's Pinacate desert, and studies of radio-tagged animals continue to support an international management plan for the antelope now being developed.

2. Conservation and Management of Parks and Forests

a. National wildlife refuges

The USFWS manages eight national wildlife refuges, totaling over 1.2 million acres, along the U.S.-Mexican border. These refuges provide for the conservation of a great variety of unique plant and animal species, including threatened and endangered species, game species, and migratory birds. They also protect many unique (including some very rare) areas of biodiversity found nowhere else in the United States or Mexico in such natural conditions. Nearly the entire border area serves as a migratory stopover or wintering habitat for numerous neotropical birds.

The USFWS also has several national fish hatcheries, fishery assistance field stations, ecological services offices, and law enforcement offices in the border states.

A number of international initiatives are underway in the border region to conserve important natural resources. In 1993, the Mexican Government declared the Alto Golfo Reserve in Sonora and Baja California, which incorporates the delta of the Colorado River and the Pinacate Desert to the east. The reserve is located along the border and opposite Cabeza Prieta National Wildlife Refuge and Organ Pipe Cactus National Monument. An international working group has been formed to coordinate management.

U.S. National Park Service staff from Big Bend National Park in Texas are working with their counterparts across the border in Chihuahua and Coahuila to explore the creation of a companion reserve there. Finally, the USFWS and U.S. National Park Service promote the "Cultural Heritage Corridor" in the lower Rio Grande River valley in Texas, where a number of national wildlife refuges are located, along with their Mexican counterparts in Tamaulitas. Implementation of the NAFTA could provide these initiatives with added impetus.

b. Habitat conservation and biodiversity

The United States and Mexico are involved in many conservation efforts to protect fish and wildlife habitat, particularly for migratory birds, threatened and endangered species, and fish. The United States is placing increasing emphasis on protection of unique and important habitats such as wetlands, and bio-diverse ecosystems.

In 1993, Mexico is taking steps to establish a Mexican Conservation Fund to be used to finance environmental programs and projects in Mexico. The purpose of the fund is to provide long-term sustained financing for building the capacity of Mexican agencies to conserve the country's biological resources and natural ecosystems. The fund will help support continuous identification of biodiversity conservation priorities and to monitor the condition and trends of natural vegetation and ecosystems. (This fund is discussed in more detail in Section IV.D.1.)

In 1988, the U.S. Congress established the San Pedro Riparian National Conservation Area, which begins at the international border and traverses the San Pedro River north for about 36 miles. This corridor is being managed by the U.S. Bureau of Land Management for its riparian, wildlife, cultural, and other natural resource values. This area has its headwaters in Mexico. In fact, about 17 percent of the San Pedro River watershed is in Mexico. This makes coordination with Mexico extremely important to the viability of the river downstream in the United States.

In the last several years, Mexico has expressed increased interest in protecting the headwaters of the San Pedro River. The Centro Ecologico de Sonora is in the process of preparing a report and recommendations to the State of Sonora and the Mexican government that are expected to suggest specific management prescriptions for portions of the watershed in an effort to protect its natural resources. The U.S. Bureau of Land Management has been asked to review the plan and provide technical assistance on the hydrologic aspects of the watershed and on other natural resources.

Following are several additional examples of recent joint habitat conservation and biodiversity efforts:

- The USFWS plays a key role in "Partners in Flight," a new conservation initiative to benefit neotropical migratory birds. Conservation projects in Mexico are now being developed under this program. USFWS personnel are active in developing guidelines for U.S.-based projects associated with Mexico.

- The loss of nesting and roosting habitat has resulted in a reduced number of breeding white-winged doves in northern Mexico. SEDESOL and the USFWS have agreed to cooperate on the management of this species. The major colonies have been cataloged and those most in need of protection identified. The Mexican ecological reserve "Parras de la Fuente" was recently established in Tamaulipas to provide long-term protection for the most important breeding colony of white-winged doves. Currently, joint efforts are underway to develop a strategy for the conservation and management of this species.

- In July 1993, the United States and Mexico agreed on several joint wetland conservation projects under the North American Wetlands Conservation Act. Under this Act, at least 50 percent of the available monies from the North American Wetlands Conservation Fund must be spent in Canada or Mexico because of the importance of the breeding and wintering habitat of the migratory birds in those countries. This supports the Tripartite Agreement on Wetlands Conservation between the United States, Canada, and Mexico. In Mexico, projects include restoration of the Chuburna Harbor dike in the State of Yucatan that was destroyed by Hurricane Gilbert. In the State of Sonora, a grant will be used to evaluate wetlands, and for research and education.

3. Conservation of Marine Resources

a. Dolphin surveys

On July 28, 1993, scientists from the National Marine Fisheries Service ("NMFS"), Southwest Fisheries Science Center ("SWFSC") and Mexico's Ministry of Fisheries ("SEPESCA") began a cooperative survey of a 606,700 square-mile area south from the Oregon-California border to Manzanillo, Mexico, extending approximately 300 nautical miles offshore. The primary objective of this joint survey is to obtain a solid estimate of abundance for the northern stock of common dolphin. Scientists will also collect data on large whales and other marine mammals, marine turtles, seabirds, and other marine life, in addition to oceanographic data such as temperature and salinity. This information will be used to describe the types of habitats that are associated with marine mammals.

b. Dolphin-safe research

In keeping with U.S. policy to develop alternative methods of fishing, NMFS plans to convene a workshop to assess and prioritize the next generation of dolphin-safe research initiatives. The objective of the workshop will be to quantify the probability of success of available detection technologies, and to develop from these probabilities a hierarchy of research priorities.

The workshop is tentatively scheduled for spring 1994. NMFS will seek the participation of Mexican scientists to aid in discussions and evaluations of proposed detection technologies. Funding for participation by Mexican scientists will be provided by the United States.

In addition, a cooperative research project initiated in 1992 to simultaneously track tuna and dolphins captured together will continue during 1993. The primary objective of the project is to study the association of these two species. This information will prove valuable in two ways. On the scientific side, the study of tuna and dolphin movements and interactions, in conjunction with food-habit studies, will help to establish the longevity and dynamics of the tuna-dolphin bond and the degree to which it is food-based. On the tactical side, the study may determine whether the bond loosens at particular times or under certain conditions, and if so, whether the tuna would be vulnerable to fishing at such times.

The study will be conducted by the Inter-American Tropical Tuna Commission, in cooperation with NMFS, the University of Hawaii, and Mexico's national tuna protection and improvement program, the PNAADP. The PNAADP will continue with research on separation of dolphins and tuna.

c. Sea lions

Although much is known about sea lions in the United States (through census, tagging, and studies of their diet, foraging and reproductive behavior), less information is available for animals in Mexico. Beginning in 1992, sea lions in western Baja California, Mexico, were censused cooperatively by NMFS and SEPESCA. Information on sea lions (specifically counts of pups during the end of the breeding season) from western Baja California is vital for making status determinations of this species.

In 1993, four census counts are planned. It is hoped that cooperative branding/marking studies will be initiated so that rates of exchange between Mexico and the United States can be determined. In addition to California sea lions, elephant seals, harbor seals, and Guadelupe fur seals will also be censused. The next survey is scheduled to begin on November 2, 1993. The use of aerial photography to collaborate this census is an important area that has recently been incorporated into the joint effort between the United States and Mexico.

d. Gray whales

Although the gray whale is no longer considered in danger of extinction, it will remain subject to prohibitions against takes under the Marine Mammal Protection Act. Also, as specified under the Endangered Species Act, populations removed from the listing must be monitored for a minimum of five years after delisting.

As a result of a decrease in pregnancy rates in the aboriginal fishery in Russia and low calf counts in Mexican waters during 1990, NMFS is examining the possibility of a joint U.S.-Mexican study to examine patterns of calf production in Mexican and U.S. waters during the winters of 1993-94 and 1994-95.

e. Small and large pelagic fisheries

The United States and Mexico have completed a joint report describing U.S. anglers in Mexican waters. In addition, a Binational Sardine Meeting was convened in Rosarito, Mexico in May 1993. Other ongoing work has emphasized the importance of exchanging fisheries data used for stock assessment. Planning is underway on the sardine biomass estimation to be conducted in U.S.-Mexican waters from Monterey, California, to Cedros Island, Baja California. This will be a joint research project among scientists from the SWFSC, the California Department of Fish and Game, and Mexico.

Several other joint research projects on large pelagics were undertaken during 1992-93, including:

- a collaborative study estimating biological parameters for swordfish;
- a joint study to compare the usefulness of various structures used in albacore age determinations;
- a joint experiment involving several Mexican scientific agencies to define the diet, vertical and horizontal movements of striped marlin, which is continuing;
- a cooperative billfish tagging program, which has resulted in over 14,000 billfish being tagged in the Baja area with the cooperation of fishing reports, charter boats and individual fishermen;
- an annual survey to collect catch and effort data for marlins fished by recreational anglers off the west coast of Mexico;
- a joint publication in English and Spanish reviewing striped marlin, swordfish, and sailfish fisheries and resource management by Mexico and the United States; and
- participation of Mexican scientists in a shark tagging and population indexing cruise off southern California.

f. Sea turtles

Mexico has implemented much stricter laws and regulations to conserve its sea turtle resources. A Mexican Presidential proclamation of May 1990 banned the taking of and trading in sea turtles. In early 1993, Mexico required that its shrimp trawl fleets in the Gulf of Mexico (numbering 300-400 boats) use turtle excluder devices ("TEDs"). With this latest action, Mexico is headed toward full, 100 percent use of TEDs well before May 1994, when U.S. law would limit shrimp imports from nations not using TEDs.

In addition, Mexico has for many years implemented programs that protect sea turtle nests on various beaches. A Presidential decree set aside 17 nesting beaches as reserve and shelter zones for marine turtles. With the establishment of a National Marine Turtle Program,

Mexico selected protection camps (beach sites) in both the Gulf and Pacific coasts where turtle eggs were studied, and protected through removal and relocation in special secure areas. In addition, this program has sought to promote turtle conservation through: the protecting of nests, fostering of artificial incubation of eggs where necessary, release of turtle hatchlings, and tagging of specimens.

D. FINANCING ENVIRONMENTAL AND CONSERVATION PROGRAMS

Environmental protection programs cost money, which often means public funds that are subject to heavy demands for many needs, particularly in developing countries. Thus, efforts to protect and improve the environment are best served by targeting domestic and international financial mechanisms directly toward environmental infrastructure projects or major conservation programs. The Border Environment Cooperation Agreement recently concluded with Mexico, establishing the Border Environment Cooperation Commission and the North American Development Bank, represents one such environmental financing mechanism directly related to the NAFTA and the U.S.-Mexico border area. (That Agreement is described in Section II.C, above.) Other national and international sources of environmental financing include debt-for-nature swaps and multilateral development bank programs.

1. Debt-for-Nature Programs

Debt-for-nature swaps present a significant opportunity to channel additional resources into protecting Mexico's environment. Through the work of a range of environmental organizations, debt-for-nature swaps have made an important contribution to conservation and environmental protection efforts in Latin America. Since 1987, $136 million in commercial bank debt has been converted through debt-for-nature swaps, channeling approximately $90 million to the environment in ten countries in the region. U.S. environmental groups continue to pursue opportunities for debt-for-nature swaps with the Government of Mexico.

Building upon a recent grant to assist these efforts, the United States recently announced that it will commit up to $20 million to help capitalize a new Mexican Conservation Fund (Fondo Mexicano para la Conservation de la Naturaleza), in an effort to preserve Mexico's forests, watersheds and biodiversity.

A consultative committee of prominent Mexican business people, academics, and conservationists has helped to design the Mexican Conservation Fund. The objective of the Fund is to provide long-term, sustained financing to strengthen the capacity of Mexican agencies and non-governmental organizations for conserving the country's biological resources and protected areas. Specific activities will be selected after broad public consultation, and will be implemented at the grass-roots level.

It is expected that the U.S. contribution to the Fund will leverage participation of other sources of finance such as the World Bank, the Inter-American Development Bank, and members of the international donor community.

Mexico has leveraged additional resources to protect its environment through debt-for-nature swaps facilitated by the Inter-American Development Bank ("IDB"). Through the IDB's program to use environmental project loans to purchase commercial bank debt, Mexico purchased $100 million in commercial debt in exchange for a commitment of local currency to a conservation area outside of Mexico City.

2. Multilateral Development Bank Activity

The World Bank, its private-sector affiliate, the International Finance Corporation, and the Inter-American Development Bank have been actively providing development finance assistance to Mexico for many years. Cumulative lending from these institutions in Mexico totals over $30 billion for some 313 projects. In the past several years, the multilateral development banks have placed increasing emphasis upon environmental considerations to promote new types of development programs designed to have beneficial impacts upon the environment. This intensified emphasis on environmental considerations is a result of U.S. policy initiatives.

a. World Bank activities

On September 28, 1993, the World Bank and the Government of Mexico agreed to a new environmental program in which the World Bank will provide up to $1.8 billion in loans, to be approved by the Bank's Board of Executive Directors, to support environmental programs in Mexico over the next three years (1994-1996). Counterpart funding from the Government of Mexico will raise the total program to $3-4 billion. The program will channel resources to strengthening Mexico's regulatory and enforcement capabilities, while addressing toxic and solid waste, natural resource management, industrial pollution, air contamination, and water supply and sanitation.

This program, which represents a significant increase in the level of funding for environmental protection and clean-up, will support two sets of initiatives. First, assistance will be provided to various regions in Mexico where integrated, comprehensive efforts addressing a broad range of environmental problems are needed.

Because of its unique problems caused by rapid urban growth and industrialization, one of the first areas to receive attention will be the U.S. border region. Indeed, the initial World Bank project will concentrate on the 14 Sister Cities along the border and will support investment in environmental infrastructure. The project aims to improve water supply and sanitation coverage through investment in new facilities and the expansion and rehabilitation of existing infrastructure. The project will also finance the closing of illegal solid waste dump sites, the construction of solid waste recycling and disposal facilities, and the training and equipping of environmental agencies at the federal, state and municipal levels to ensure compliance with Mexico's environmental standards. To finance these efforts, the World Bank is to provide approximately $350 million, with the Mexican government contributing an additional $350 million.

The second set of initiatives under the World Bank loan will focus on nationwide strategies and will be aimed at various sectors to be improved, including water supply, waste management and biodiversity protection. Two approaches, one regional and one national in scope, will be used in tandem to maximize local participation in project design and implementation while strengthening Mexico's national environmental protection capabilities.

Collaboration between Mexico and the World Bank in environmental activities dates back to 1973. In that year, a $90 million Mexico City Water Supply loan was signed; in subsequent years, Mexico has received over $400 million in World Bank loans for water treatment and sanitation. More recently, in April 1992, the Bank supported a $50 million effort to upgrade the capabilities of the two leading environmental agencies at the federal level -- the technically-oriented National Ecology Institute ("INE") and the enforcement-oriented Federal Attorney General for Environmental Protection ("PFPA"). In December 1992 Mexico and the Bank initiated a $220 million Transportation Air Pollution project for the Mexico City metropolitan area. This project financed the replacement and retrofitting of public transport vehicles to meet stringent emissions standards, investments to improve gasoline quality and reduce fuel evaporation, and improvement of air quality monitoring and planning. The lessons learned from these and other projects will be incorporated in the new environmental program.

b. International Finance Corporation activities

The International Finance Corporation ("IFC") is also applying its experience in environmental infrastructure financing in Mexico.

The IFC Board of Directors may soon consider an investment in a privately-owned wastewater treatment facility in Puerto Vallarta. This transaction would serve as a model for "BOT" (build-operate-transfer) infrastructure financings in Mexico including, perhaps in the not-to-distant future, two wastewater treatment facilities to be constructed in Ciudad Juarez.

In addition to infrastructure, the IFC is working with the private sector in other environmentally-beneficial ways; last year, for example, it helped to finance a feasibility study for the consolidation, collection, and incineration of hospital wastes in Mexico City.

c. Inter-American Development Bank activities

The Inter-American Development Bank ("IDB") is also heavily involved in environmental infrastructure financing in the border region, most specifically the Tijuana wastewater project included among the Sister Cities projects. In this instance, a $46.4 million IDB loan to extend sewer system coverage, approved in 1985, was modified to take into consideration the binational, IBWC-sponsored sewage treatment facility in San Diego. The IDB has also financed wastewater systems in Monterrey.

The IDB has three relevant global project loans in Mexico which are currently being disbursed: $200 million for irrigation and drainage; a $300 million project (co-financed with the World Bank) for potable water and sewage; and $350 million for energy conservation. A follow-up global irrigation and drainage project ($300 million), as well as a sewage project in Guadalajara ($100 million) and a basic sanitation project for Puebla ($100 million) are also in the IDB's pipeline.

Other IDB resources that may become available in the near term include a $180 million municipal development project (to be matched by Mexican government resources) to improve the administration and operation of municipal services throughout the country. A mix of credits and grants will be aimed at developing local infrastructure and reinforcing institutional capacities, including establishing a fee-for-service basis for municipal services such as clean water and sewage. Board approval is anticipated in early 1994.

The IDB currently provides approximately $1 billion in new loans to Mexico each year. Additional IDB efforts to enhance the environment in Mexico may be in the offing, prepared in the context of the current IDB capital replenishment negotiations.

3. The U.S.-Mexico Border Infrastructure Finance Conference

Rapid economic and population growth along the U.S.- Mexico border has outpaced the development of necessary environmental infrastructure. This "infrastructure deficit" has strained the environmental balance in the border area and resulted in serious problems in such areas as water pollution and waste disposal.

As part of the effort to respond to these problems, the U.S. Department of Commerce and SEDESOL convened the U.S.-Mexico Border Infrastructure Finance Conference in July 1993 in San Antonio, Texas. Conference participants included seven cabinet-level representatives from the U.S. and Mexican governments.

The conference explored methods of attracting private capital to finance border environmental infrastructure projects. The conference also examined border infrastructure needs, challenges to financing these projects, and potential solutions to the border environmental infrastructure deficit.

Conference participants explored the possibility of obtaining private financing for border environmental projects in the water supply, wastewater treatment, municipal waste, and hazardous waste areas. Other financing issues analyzed at the conference included: the ability of the federal government to at least partially fund environmental infrastructure when the primary beneficiaries of the projects are border citizens and communities, rather than private industry; the need for the development of user fee mechanisms for environmental infrastructure projects; and the ability to finance environmental projects that benefit private users, such as industrial wastewater projects.

The conference discussions contributed to the development of proposals for a mechanism based on the successful financing experience of development banks, which became the basis for the recently-concluded agreement between the United States and Mexico to establish the North American Development Bank. This bank will provide some $2 billion or more of largely private market financing for environmental infrastructure projects in the border area.

V. OVERVIEW OF POTENTIAL ENVIRONMENTAL EFFECTS OF NAFTA

A. MACROECONOMIC EFFECTS

1. Effects on Economic Growth

Implementation of the NAFTA is expected to promote additional development of both the U.S. and Mexican economies. In the short run, this economic development will take place through the increased specialization and economies of scale resulting from the freer flow of trade in goods and services. In the long run, however, the expansion of research and development, training, and capital equipment encouraged by the NAFTA will enhance economic activity even more. For Mexico, improved access to U.S. technology in the energy, communications, electronics, and other critical industries should provide an additional boost to growth.

Defeat of the NAFTA, on the other hand, would not perpetuate the status quo. Rather, it would be almost certain to reduce U.S. exports to Mexico. Mexico could suffer from a loss of confidence in its economic growth prospects, reduced domestic and foreign investment, and slower growth. While Mexico would surely exercise policy options in an attempt to preserve its prospects for economic development, these would be limited, at least in the short run. The near-term effects of NAFTA's defeat would result in less growth and more poverty in Mexico, and in the United States, would result in reduced exports to Mexico, fewer high-paying U.S. jobs supported by those exports, and ultimately, higher levels of immigration from Mexico than would otherwise be the case.

The impact of the NAFTA on economic growth could be substantial. It is estimated that, NAFTA would cause:

- Mexican economic activity to be as much as 11 percent greater;
- Mexican per capita incomes to rise even more if higher incomes cause birth rates to fall, as expected; and
- U.S. exports of goods and services to Mexico to increase substantially, increasing U.S. jobs supported by merchandise exports by an additional 200,000 by 1995 by almost 40 percent, bringing over 350,000 gross new jobs to the economy.

Most serious economic studies of the NAFTA suggest that it would boost U.S. GDP by as much as a half percent. Estimates of net employment gains range from 35,000 to 170,000. Although the additional output and jobs are just a small fraction of the growth that normally occurs each year in the U.S. economy, they will be especially important during periods of slow economic growth (Figure 1) and in the states bordering Mexico. For instance:

FIGURE 1

EXPORT GROWTH SINCE 1987
HAS SIGNIFICANTLY BOOSTED
U.S. ECONOMIC ACTIVITY

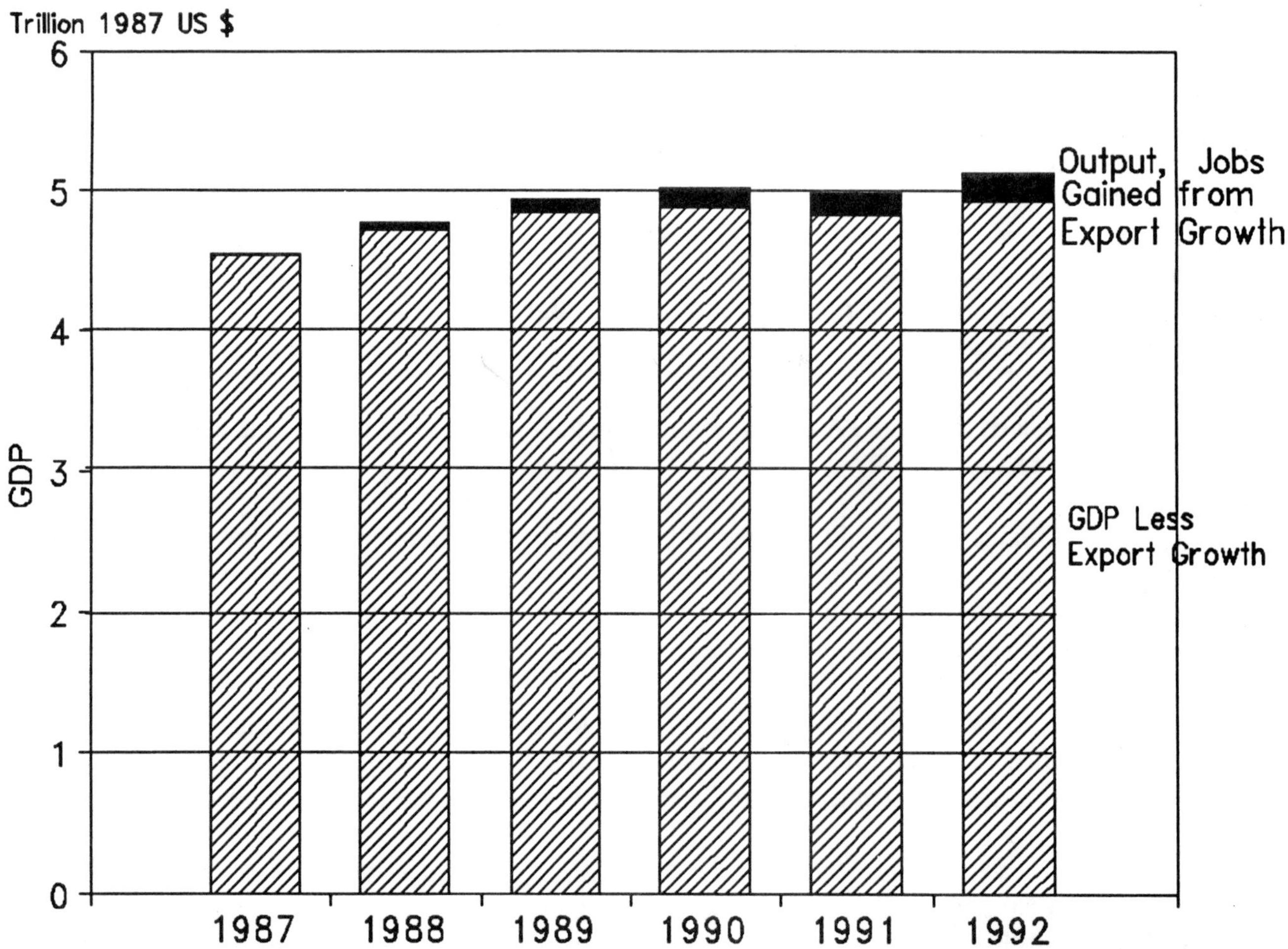

- Increased exports to Mexico added an estimated 180,000 jobs to the U.S. economy from 1990 to 1992, helping to soften the recession. As a result of the recession, civilian employment in 1992 was down 300,000 compared to 1990.

- Of the 266,000 jobs created by the growth of manufactured exports to Mexico from 1987 to 1992, Texas captured 31 percent, California 17 percent, and Arizona and New Mexico 3 percent combined (Figure 2).

The NAFTA is expected to have a relatively larger impact on the Mexican economy, which is only one-twentieth the size of the U.S. economy. Most economic analyses suggest the NAFTA will add perhaps one percentage point to Mexico's annual economic growth rate. Most of this additional growth is expected to occur in the industrial and service sectors. Unlike the maquiladora program, however, which is concentrated largely in northern Mexico near the U.S. border, NAFTA-related growth will probably be more evenly distributed throughout the country, focusing on major Mexican consumer and investment markets (although it will also include some growth in border areas).

Mexico's accelerated growth will depend greatly on increased investment, especially from the United States and other foreign countries. From 1987 to 1991, U.S. direct investment in Mexico averaged about $2 billion dollars a year, accounting for two-thirds of all such investment in Mexico. At the same time, it accounted for only six percent of total U.S. direct investment abroad and was less than 0.3 percent of total investment in the U.S. economy during that period. The NAFTA is expected to increase U.S. investment in Mexico. But even were it to increase substantially, there would be little impact on investment in the United States, as recently confirmed by the Congressional Budget Office[4] (Figure 3).

The increase in foreign investment stimulated by the NAFTA will make it possible for Mexico to significantly increase its imports. Most economists expect Mexico to allow its foreign trade deficit to balloon for at least the next decade or two as it industrializes. Because of proximity and growing direct investment in Mexico, the United States could greatly benefit from any expansion of imports. In 1992, Mexico used almost 17 percent of its income to buy goods and services from the United States. Under NAFTA, this share is likely to rise toward the Canadian level of 21 percent, as Mexican industries increase their reliance on U.S. capital, inputs, business, engineering, financial services, and also environmental technology and waste management services. Rising wages will also increase Mexican demand for U.S. consumer products, already one of the fastest growing components of U.S. sales to Mexico.

a. The environmental consequences

Like most advanced economies in the course of development, Mexican industrialization will increase environmental pressures. Mexico already is encumbered with serious industrial pollution problems in its major cities and in its cities along the U.S. border. The NAFTA

---

[4] Congressional Budget Office, "A Budgetary and Economic Analysis of the NAFTA," July 1993.

FIGURE 2

U.S. EXPORTS OF MANUFACTURED GOODS TO MEXICO:
STATE SHARE OF JOBS GENERATED BY EXPORT GROWTH
1987-92

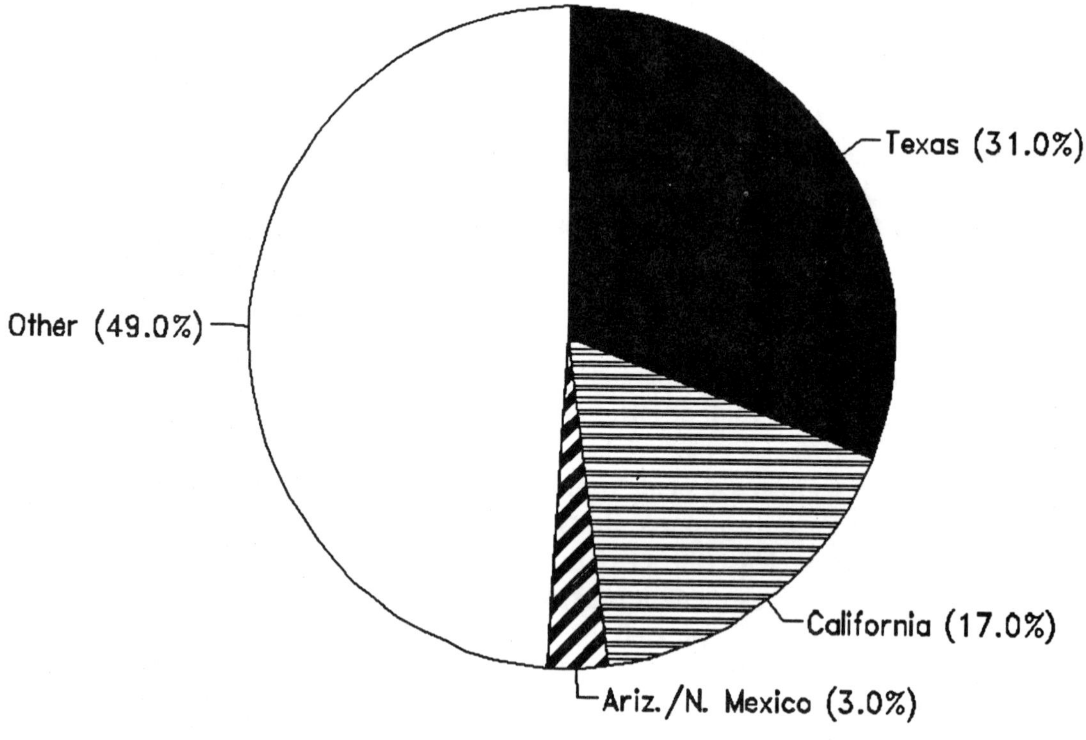

Total Job Gains from Growth of Manufactured Exports:
266,000

FIGURE 3

U.S. INVESTMENT IN MEXICO, TOTAL
DIRECT FOREIGN, AND DOMESTIC
1992

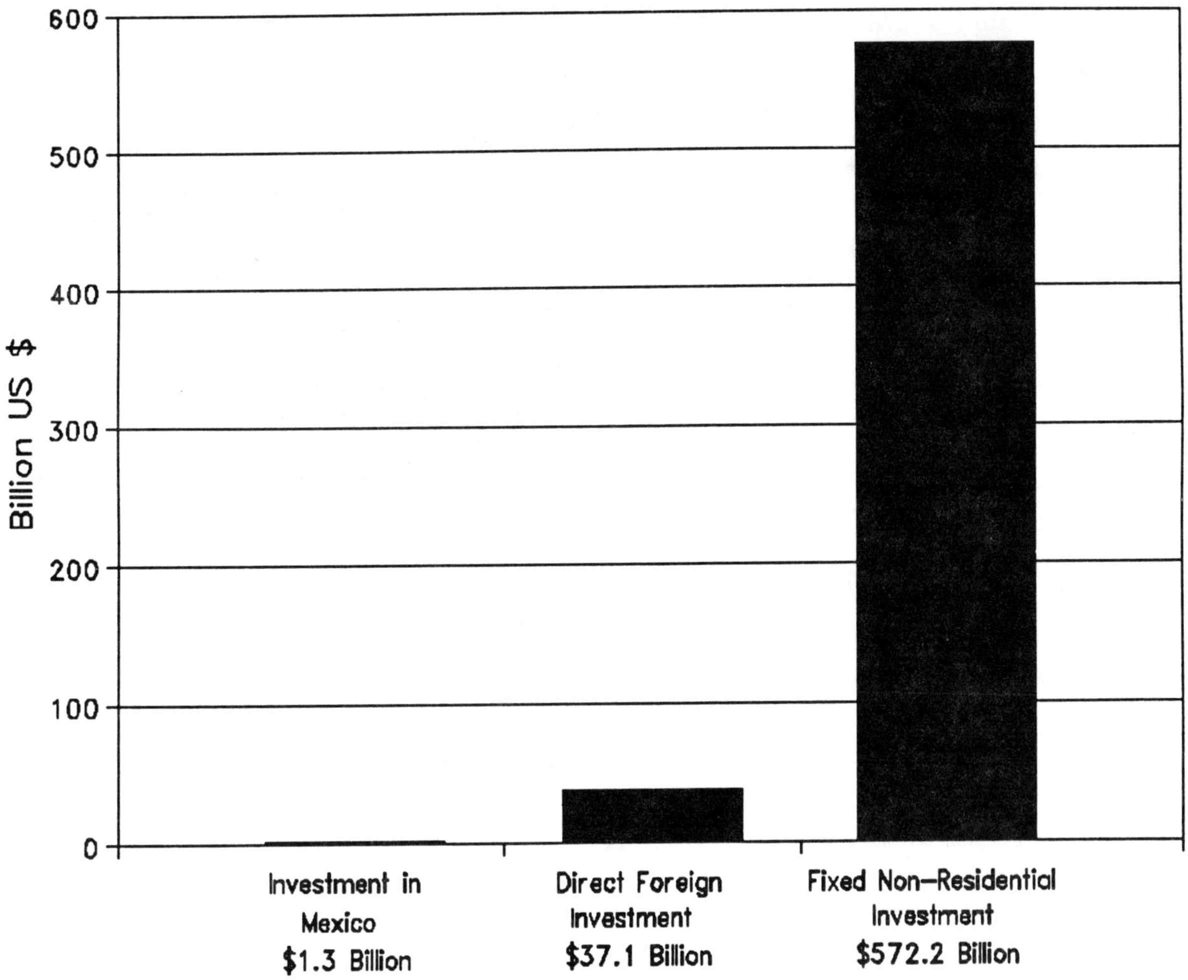

will tend to aggravate problems associated with Mexican industrial centers and transport arteries. Where those centers and arteries lie close to the U.S. border, some of the environmental effects may spill over into the United States. Otherwise, the economic effect of these problems on the United States will be positive, to the extent that Mexico seeks U.S. goods and services to deal with them.

The NAFTA could also create additional environmental pressures in those border states most likely to benefit from increased economic activity due to NAFTA. Because most benefits to the United States are small in comparison to total economic activity, the direct impact on U.S. environmental problems is likely to be minimal. The additional economic activity in the United States resulting from the NAFTA should pose no substantial increase in the demand for environmental services and funds, or investment funds in general. This may be less true, however, for Texas and California, where a large part of the NAFTA-induced economic activity is likely to take place.

At the same time, the NAFTA is likely to greatly facilitate Mexican efforts to cope with development-associated environmental problems:

- Increased integration of the North American economy will make the latest, most environmentally sound technology available to Mexican industries, allowing them to bypass dirtier methods used by other countries in industrializing.

- The NAFTA Environmental Agreement and the Border Environmental Cooperation Agreement offer Mexicans the opportunity to confer with U.S. agencies and enlist U.S. help in dealing with environmental problems. At the same time, under NAFTA and the Environmental Agreement, Mexico has committed not to relax its environmental standards or enforcement to attract foreign investment.

- The NAFTA will encourage a wider distribution of industry throughout Mexico. While this may bring some pollution to new areas, it is also likely to slow the growth of pollution in areas already severely affected, making it easier to implement effective programs to remedy environmental problems in those areas.

By speeding up Mexican industrialization and associated wage increases, the NAFTA will spur the growth of per capita income. Economic studies have shown that societies are unwilling or unable to devote resources to environmental problems as long as income levels are very low. Mexico is no different. The faster its per capita income rises, the sooner and more effectively it can deal with its environmental problems, including those that spill over into the United States.

As Mexican per capita incomes rise :

- Population growth rates are likely to fall, reducing population growth pressures and further accelerating the rise of living standards (Figure 4);

- The availability of safe water and urban sanitation will spread, reducing mortality rates and the incidence of disease and morbidity; and

FIGURE 4

RISING INCOME ENCOURAGES LOWER BIRTH RATES

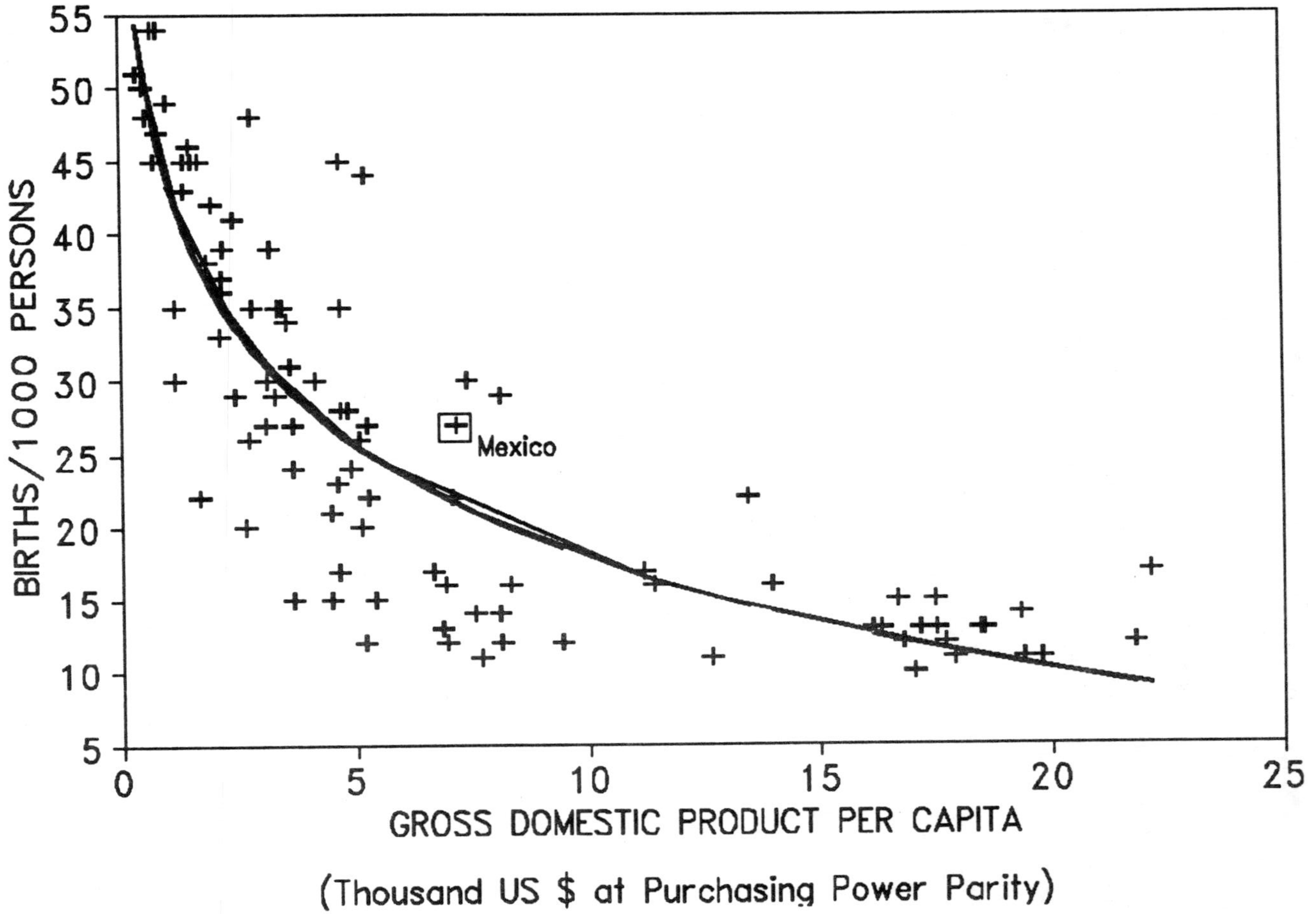

- Concentrations of particulate matter will decline.

Urban concentrations of sulfur dioxide tend to peak when per capita incomes reach several thousand dollars and drop off rapidly thereafter.[5] Mexico's per capita income has now reached this level, leading environmental experts to believe that by accelerating the rate of income growth, as the NAFTA is likely to do, Mexican problems related to this particular pollutant can be reduced more rapidly.

2. Investment Implications

Available evidence and empirical research suggest that environmental considerations are generally not important determinants of investment decisions in North America. Furthermore, under NAFTA and the Environmental Agreement, each country has committed to avoid environmental policies intended to divert investment from one country to another. Finally, NAFTA will remove the current artificial incentives which have intensified investment along the border through the maquiladora program.

Without NAFTA, it is more likely that intense border investment will continue, with the attendant adverse environmental consequences for the border region.

Evidence suggests that environmental considerations are not important determinants of investment decisions in North America. Economic theory does suggest that a country can capture a competitive advantage by being more lax than its trading partners in addressing environmental problems. In practice, however, this edge may be partially or entirely offset by other factors -- such as other production costs, relocation costs, tariffs, and tax considerations.

Cross-country studies of trade patterns have found little evidence that environmental protection costs significantly influence investment decisions and trade patterns.[6] Looking specifically at U.S.-Mexican trade, Grossman and Krueger[7] examined the influence of abatement costs in U.S. industries as a share of the total valued added in each industry. They found no significant relationship between these costs and:

---

[5] Gene M. Grossman and Alan B. Krueger, "Environmental Impacts of a North American Free Trade Agreement," paper prepared for a conference on U.S.-Mexico Free Trade Agreement, Brown University, 3 October 1991.

[6] Ingo Walter, "Environmentally Induced Industrial Relocation to Developing Countries," in S.J. Rubin and T.R. Graham (eds.), Environment and Trade (New Jersey: Allanheld, Osmun, and Co., 1982); H. Jeffrey Leonard, Pollution and the Struggle for World Product (Cambridge: Cambridge University Press, 1988); and James A. Tobey, "The Effects of Domestic Environmental Policies on Patterns of World Trade: An Empirical Test," Kyklos, 43:191-209.

[7] Op. Cit.

- the propensity of specific industries to import from Mexico;
- imports from Mexico with a high content of U.S.-made components; or
- the ratio of maquiladora value-added to U.S. value-added in specific industries.

Grossman and Krueger attributed the absence of any evidence of a Mexican trading advantage related to environmental costs to the fact that pollution abatement costs are a very small fraction of the total costs incurred by most industries, averaging only 1.4 percent across all U.S. manufacturing industries.

A field study conducted by the Harvard Business School[8] suggests additional reasons. Mexican environmental regulations in principle are not significantly different from those in the United States. Regulations require new plants to use the best available technology and specify nine "critical areas." While less stringent standards are applied to existing plants, some existing facilities are required to attain the same standard as applied to new sources (although they are given more time to come into compliance).

Although Mexican enforcement is not as extensive as it is in the United States, the Harvard study found that Mexico can credibly monitor new investments for environmental compliance. Mexico's per capita spending on environmental enforcement climbed from 8 cents in 1989 to 48 cents in 1991. Comparing compliance, inspection, and enforcement staff to gross industrial product suggests that Mexican efforts are comparable to those in neighboring U.S. states (Figure 5). In addition, Mexico's strategy of concentrating enforcement efforts on highly polluting urban facilities may be a very effective initial step toward significant reductions in emissions.

Moreover, the Harvard Study found that U.S. company perceptions did not support fears of widespread relocation to Mexico. A survey of companies in industries with relatively high pollution abatement costs revealed that:

- Managers of these companies believed that the gap between Mexican and U.S. costs is closing and will be narrower in the future.
- In addition, while the managers perceived Mexican costs as lower than those in various U.S. regions, they also believed that Mexican costs are higher than those in Southeast Asia, South America, and Africa.

Under NAFTA Article 1114 and the Environmental Agreement, each NAFTA country has committed to avoid environmental policies intended to divert investment from one country to another. If such policies can be avoided, the evidence suggests that existing and expected environmental conditions are not likely to be important factors in U.S. company investment decision-making. The Harvard Business School study concluded that:

---

[8] Reported to the Environmental Protection Agency by Todd Collins, Richard Hojel, Daniel Jinich, John Lonnquist of the Harvard Business School 11, May 1992.

FIGURE 5

A COMPARISON OF U.S. AND MEXICAN ENVIRONMENTAL ENFORCEMENT EFFORTS

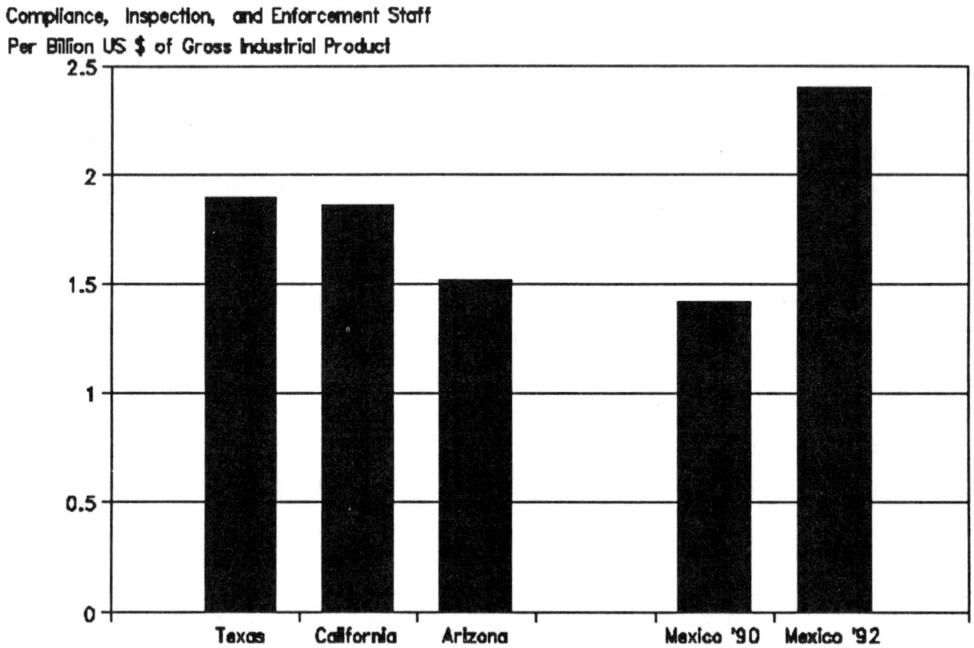

* Compliance, Inspection, and Enforcement (CIE) staff includes all front-line personnel involved in monitoring and visiting plants. For the United States, data was compiled from EPA Regional Offices and state agencies, with state agencies providing estimates of additional local CIE resources.

* Gross Industrial Product (GIP) includes the following sectors of Gross National Product (GNP): mining, construction, manufacturing (durable and non-durable), transportation, communications, and public utilities.

- Pollution intensity and control costs are not good predictors of decisions to invest in Mexico; and

- Mexican enforcement strategy will effectively control traditional pollution-intensive industries.

The study found no clear relationship between the level of industry pollution abatement costs and reported management interest in Mexican investment. Only 10 percent of the responses mentioned pollution abatement costs as being among the two most important advantages of Mexico in determining investment location.

The Harvard group supported its survey data with a case study of businesses in Los Angeles. The case study found that:

- A 1987 regulation requiring a 93 percent reduction in VOCs in 10 years was viewed by Los Angeles' 1200 furniture makers as a "shutdown order." Because furniture making is compatible with Mexican resources and proximity, it was feared that many of the companies would relocate across the border. But only 128 firms (11 percent) actually left Los Angeles, and only 28 of these (2 percent) relocated to Mexico.

- A subsequent survey of all local businesses by the Los Angeles Small Business Coalition found that only 6 percent were planning to leave the city, but only 15 percent of those (less than 1 percent of the total) planned to move to Mexico.

a. NAFTA's impact on Mexican maquiladoras

In recent years, the economy of the border area has been dominated by assembly plants known as maquiladoras. These plants have been set up under Mexico's maquiladora program, which was established in 1965 to attract foreign investors through special incentives, including tariff waivers and exemptions from Mexico's foreign equity restrictions. When the program was created, maquiladoras were limited to Mexico's border region. Because of the region's easy access to the United States, many investors established assembly operations designed to take advantage of U.S. provisions in the tariff schedules which assess duties only on the factory value added for goods assembled in any other country (including Mexico) using U.S. components.

Today, there are over 2,000 maquiladoras employing over 400,000 people. Most are located in the border region, though some have been established in Mexico's interior. Many maquiladoras are assembly plants, importing components duty-free from the United States and other countries and generally exporting their finished products to the United States. Some maquiladoras also export to countries other than the United States, and a few enjoy tightly restricted permission to sell a fraction of their production in Mexico. They are subject to all Mexican laws regarding the environment and labor, and must return to the United States the waste that imports generate.

As Mexico has relaxed its foreign investment restrictions in recent years, the maquiladora program's chief advantage has become its waiver of Mexican duties on imported parts and equipment. In return, maquiladoras must export all of their production unless they obtain Mexican government permission to sell a limited quantity in Mexico. They are also bound by a variety of Mexican incentives for all firms to limit imports, export more, and use Mexican materials.

NAFTA effectively eliminates the key attractions of the maquiladora program. NAFTA provisions sharply restrict export-linked duty drawback and waivers. This means that maquiladoras will pay full Mexican duties on parts imported from outside of North America and used in products sold in the United States, Canada or Mexico, beginning in 2001. NAFTA also commits Mexico to eliminate restrictions on maquiladoras' sales into Mexico's market, and other export-based performance and domestic content requirements, by 2001. Maquiladoras will operate the same way as other factories in terms of paying tariffs on imported parts and freedom to sell without restriction in Mexico. Maquiladoras will also continue to be required to return their hazardous waste to the United States.

NAFTA provisions will not eliminate maquiladoras. It will require the maquiladoras to play by the same rules as other firms with respect to export incentives. This may result in fewer new firms constructed in the border area.

B. SECTORAL EFFECTS

1. Energy

Concern for a more efficient, less vulnerable, and environmentally sustainable energy future was the foundation of the negotiations of the energy provisions of NAFTA. By removing barriers and disincentives to the use of natural gas in Mexico, NAFTA opens up prospects for cleaner power generation. This, together with greater scope for private investment in Mexico's power generation sector, should help mitigate environmental effects of increased demands for electricity created by economic growth.

By aligning Mexican oil and gas prices with market forces, NAFTA will open greater opportunities for access to renewable energy technologies. NAFTA also allows continuation of incentives for exploration and development of non-renewable energy sources, but does not change disciplines on subsidies spelled out elsewhere in the Agreement.

Absent NAFTA, the benefits of greater use of natural gas and of cleaner technologies could be lost. If the NAFTA is not approved, U.S. opportunities to market goods and services in the Mexican economy would be substantially reduced and the environmental benefits from increased sales of U.S. energy goods and services, including natural gas and renewable energy technologies, would be significantly curtailed.

The United States and Mexico are major producers and consumers of energy. Trade in energy products is a significant part of the economic relationship between the two countries. Fuels alone represent about 19 percent of U.S. imports from Mexico. Nearly half (42 percent) of the crude oil stored in the U.S. Strategic Petroleum Reserve was supplied by Mexico. The United States is a major supplier of petroleum product imports to Mexico, providing over 75 percent of its imports.

Mexico officially estimates its crude oil reserves to be twice those of the United States. These reserves represent nearly half of the non-OPEC, global reserves located outside the countries of the former USSR. Most of these reserves lie in offshore areas along the coast of Mexico. U.S. firms are at the frontier of a number of new exploration and production technologies, and much of the experience with these technologies resides exclusively with U.S. firms.

Mexican crude production has increased very little since 1982. Given current financial constraints, it is questionable whether Mexico can mobilize the large amount of capital needed to substantially increase production from current levels. Development of Mexico's offshore oil reserves is not expected to be affected significantly by NAFTA, since private investment in such activities is still not permitted under Mexico's constitution.

The provisions of NAFTA generally liberalize other aspects of Mexico's state-regulated energy regime. NAFTA immediately lifts Mexican investment restrictions on 14 of the 19 basic petrochemicals previously reserved to the Mexican states, and on 66 secondary petrochemicals. Mexico now will allow the private ownership and operation of electric generating plants for self-supply, co-generation and independent power production, which could lead to more efficient use of energy resources and replacement of outmoded technology. NAFTA permits U.S. natural gas and basic petrochemical suppliers to negotiate directly with Mexican end-users for the sale of natural gas or basic petrochemicals and to conclude contracts with the buyers together with PEMEX.

In addition, Mexico's new Mining Law frees mines from government control. As new dual coal/oil and coal-fired plants come on stream, U.S. and Canadian coal will compete tariff-free in an expanded Mexican market. Opportunities are also expected for new clean coal technologies.

One of the products basic to the improvement of Mexican air quality is methyl tertiary butyl ether ("MTBE"), used as a gasoline additive to replace lead and abate carbon monoxide (CO). The Mexican government is increasing the availability of unleaded gasoline. In addition, the Mexican government has removed MTBE from the list of basic petro-chemicals reserved for production by PEMEX, to encourage private sector investment in MTBE plants. MTBE was purchased from U.S. sources by PEMEX for use in Juarez for the 1992-1993 winter season, for CO abatement.

Although U.S. crude oil reserves and production have been declining since 1985, natural gas reserves remain substantial and coal is the most abundant fossil fuel in the United States. The United States leads in developing technologies that use coal in more economical and environmentally acceptable ways. The fraction of U.S. energy supplied by renewables has been increasing, particularly in areas with no sunk investment in centralized power distribution and generation.

Energy policies must be consistent with improving and maintaining air, land and water quality in Mexico and the United States. One means of doing this is greater use of natural gas, the cleanest burning fossil fuel. As pointed out in the 1992 Environmental Review, natural gas represents a significant opportunity to meet Mexico's growing energy demands while at the same time reducing emissions of harmful pollutants. It can also be used to effectively reduce emissions from existing electrical generation plants. Mexico's shift over the last decade from being a net exporter of gas to the United States to being a net importer is likely to continue over the remainder of this decade, providing a growing market for the U.S. gas industry.

NAFTA liberalizes trade barriers to natural gas to permit it to be used, not only to generate electricity in more locations, but also to supply energy to Mexico's industrial sector (which will be particularly important in enhancing Mexico's environmental quality). NAFTA also encourages more private investment in Mexico's power generation sector, which will complement all these energy sector developments.

NAFTA will allow the prices of Mexican oil and gas products sold in its domestic market to be set by market forces. By eliminating price distortions, such a shift will also allow U.S. renewable energy technologies to gain access to the Mexican markets in which they can best compete. These are most likely to be the areas farthest from existing electricity distribution systems, into which it has been uneconomic to extend the grid. The NAFTA will also provide patent protection for energy technologies by means of its provisions protecting intellectual property rights.

Finally, while NAFTA allows continuation of incentives for exploration and development of energy resources, it does not exempt such incentives from its disciplines on subsidies. This provision was carried over from the Canada-U.S. FTA.

2. Agriculture

NAFTA is expected to result in increases of U.S. exports of grains, oilseeds, and meats, while fruits and vegetables are likely to account for the bulk of Mexico's increased exports. Effects on environmental quality depend in part on resulting changes in the output and input mix, and also on land use patterns. Chemical use changes will differ regionally, but an overall net result is difficult to discern. Mexico's corn market liberalization, and reduction in water and chemical input subsidies, could result in reduced cultivation of marginal lands, reducing potential for erosion, deforestation and loss of biodiversity.

For the United States and Canada, the agricultural environmental consequences of not concluding the agreement are small. For Mexico, however, two recent unilateral changes -- changes in current land tenure laws and the shift from corn support prices to decoupled support payments -- increase the risk of negative environmental effects if NAFTA is not concluded. Much of Mexican corn production is concentrated on small farms in rainfed areas and many of these farms are on marginal, environmentally fragile land. In addition, much of the remaining forested land is held communally. Without NAFTA and the accompanying increase in wages and jobs for landless Mexican workers, there will be an increase in the cultivation of marginal lands and deforestation of forested lands as landless workers leave corn production.

NAFTA will present agricultural producers in the United States, Canada, and Mexico with a new constellation of input and output prices. As these producers respond to the new prices, they will change their use of inputs such as pesticides, labor and land, and the level and mix of outputs. These production changes can improve or worsen water quality, soil erosion, soil productivity, biodiversity, wildlife habitat, food safety, and worker health.

NAFTA's net impact on U.S. and Canadian agricultural environmental quality will be small because price changes faced by U.S. and Canadian agricultural producers will be small. In Mexico, more substantial price changes are expected. In particular, it is expected that Mexican wage rates will increase, producers of Mexican fruits and vegetables will see an increase in their prices, Mexican livestock producers will see a decrease in feed prices, and Mexican corn and sorghum producers will face price declines.

NAFTA will also liberalize investment opportunities, with the greatest changes expected in Mexico. Changes in investment rules may encourage regional shifts in agricultural processing, which can contribute to surface and groundwater pollution. Although U.S.-Mexico border pollution has received considerable attention, agriculture is not the predominant sector at the border; current efforts to address border pollution within NAFTA will have little effect on environmental problems associated with agricultural production and processing.

Under NAFTA, exports of grains, oilseeds, and meats would account for most U.S. export expansion, while fruits and vegetables would account for a large portion of the increase in Mexican exports to the United States. Although U.S. exports of some commodities would increase, U.S. output would increase only slightly because exports to Mexico of those commodities constitute only a small portion of total U.S. crop production. On the other hand, Mexican imports and exports to the United States account for a greater percentage of Mexican agricultural production. Therefore, output adjustments for Mexican agricultural production would be proportionately larger than changes expected for U.S. agriculture.

NAFTA is not expected to produce any significant effects on the structure of American agriculture (i.e., concentration of production). NAFTA is likely to bring about some regional production effects, and a net overall expansion in U.S. agricultural exports.

Changes in environmental quality depend on changes in output and input use, not on changes in trade flows. Increased agricultural output can result in an increase in land under production, an increase in total agricultural chemical use (fertilizer and pesticides), and an increase in the intensity (per acre rate) of chemicals used. The degree to which increased land or chemical use affects resource quality depends critically on such geographic and physical characteristics as soil type, slope, weather, proximity to surface and groundwater, and habitat sensitivity. Other, less direct, environmental impacts can occur through changes in wage rates which affect land use and production patterns, and through changes in investment rules which can influence technology transfer.

In the United States, chemical input use in grain production would increase slightly, with small net increases in planted acreage expected under NAFTA. On the other hand, U.S. planted acreage of fruits and vegetables would decrease slightly, leading to reductions in chemical inputs on those crops and U.S. demand on limited water resources (although it is unlikely that this fruit and vegetable acreage would be completely idled; when possible, farmers would produce other commodities). Overall, there would be a small increase in total U.S. chemical input use.

More important from an environmental perspective are regional changes in chemical and land use: chemical use increases are expected in corn-producing states such as Illinois, Indiana, Iowa, and Nebraska. Some portions of Illinois, Indiana, and Iowa are considered highly vulnerable to groundwater contamination from pesticides and nitrates.[9] Small increases in chemical use are also expected in wheat-producing areas of Kansas and North Dakota and in sorghum-producing states such as Kansas, Nebraska and Texas. These states are considered

---

[9] Kellogg, Robert L., Margaret Maizel, and Don Goss. "Agricultural Chemical Use and Ground Water Quality: Where are the Potential Problem Areas?" U.S. Department of Agriculture, Soil Conservation Service, Economic Research Service, Cooperative State Research Service, and National Center for Resource Innovations. December 1992.

relatively less vulnerable to pesticide and nitrate leaching.[10] Chemical use will decline in Arizona, California and Florida, where fruits and vegetables are produced. Florida production regions are considered highly vulnerable to contamination of groundwater by pesticides and nitrates.[11]

In Mexico, lower corn and sorghum output will cause some areas to either come out of production, shift into pasture, or shift into production of fruits and vegetables or other crops, such as cotton. There is not adequate information to predict resulting changes in chemical use in these areas. The increase in other crop production will offset some of the declines in chemical use in corn and sorghum. Consequently, if there are environmental problems associated with chemical use and irrigation in fruit and vegetable production regions, they will continue. On the positive side, the effects of the recently announced liberalization of the Mexican corn market combined with higher wages will reduce the profitability of using marginal lands for crop production, reducing the potential for erosion, deforestation, and loss of biodiversity.

Consequently, although fruit and vegetable production is expected to increase, production of other chemical-using commodities, such as sorghum, is likely to decline. In addition, because Mexico is concurrently reducing subsidies for water and chemicals, farmers are expected to apply fewer chemicals and use water more efficiently, thus ameliorating any increase in input use due to increased output and changes in crop mix. While total use may not change significantly, what is important from an environmental perspective is the intensity of chemical use in geographic areas vulnerable to chemical leaching.

Mexican livestock production under NAFTA is likely to respond positively to lower imported feed prices and to increased access to the U.S. market, which could expand Mexican cattle (cow/calf) operations in the northern states and along the border (in areas with pasture or pasture potential). Because these operations rely on pasture for feed, currently cropped areas will potentially be converted to pasture. If input-intensive crop areas are replaced with properly managed pastureland, environmental gains can result, because pastureland is relatively less environmentally degrading. But if forestland or marginal lands are used for grazing, there could be negative environmental effects.

Canadian forestry issues also pose some special problems. One such problem is that much of the forested land in British Columbia is owned by the government, which sets stumpage fees for harvesting timber on government-owned land which the United States Department of Commerce has determined to represent an unfair subsidy to Canadian lumber. Although the NAFTA will not affect stumpage fees per se, under the NAFTA parties will still be able to impose countervailing duties to offset subsidized imports that injure a domestic industry. Also,

---

[10] Id.

[11] Pettyjohn, Wayne A., Mark Savoca, and Dale Self. "Regional Assessment of Aquifer Vulnerability and Sensitivity in the Conterminous United States." Robert S. Kerr Environmental Research Laboratory, Office of Research and Development, U.S. Environmental Protection Agency.

the Environmental Agreement's procedures for monitoring and promoting environmental quality may encourage a reassessment of pricing of forest resources.

3. Transportation

By promoting economic growth and increased trade, NAFTA will heighten transport-related environmental concerns such as congestion, noise, and emissions. However, NAFTA also contains offsetting provisions that address the same concerns, such as opening up cross-border trucking to avoid delays, and eliminating the need for inefficient "empty" return trips, and risky transfers of hazardous cargoes at the border. All trucks operating in the U.S. market will be required to meet U.S. safety standards. NAFTA also provides a vehicle for upward harmonization of safety and emission standards through the work of the Automotive Standards Council. Recently introduced efficiencies in rail links should facilitate intermodal carriage and alleviate some of the added pressures on the trucking sector.

The NAFTA also provides specific customs administration procedures to streamline import and export procedures and to ease congestion at customs border points. This will alleviate air pollution from vehicles in U.S.-Mexican border cities. NAFTA's Customs Administration provisions establish a Working Group, which will meet regularly to address ways to facilitate trade flows, including harmonization of automation requirements and documentation, and proposed administrative and operational changes.

Absent NAFTA, the environmental problems created by increasing border traffic would continue to exist, but without some of the offsetting provisions of the transport chapter of NAFTA, and without the promising work planned for transportation standards.

a. Trucking and road transport

NAFTA will liberalize trucking, rail, and bus service between the United States and Mexico, thereby decreasing border congestion in a number of ways. First, it will allow trucks and railroads to transport cargoes directly to their destination in both countries. For trucks, this means cutting idling time and ending the need to switch trailers at the border and return home empty.

Current Mexican law prohibits foreign-owned or operated motor carriers from doing business in Mexico. Retaliatory U.S. law forbids new Interstate Commerce Commission ("ICC") grants of operating authority to Mexican motor carriers wishing to operate in the United States, and limits Mexican motor carrier operations to ICC-defined commercial zones along the U.S.-Mexican border. Thus, U.S. truckers transporting goods to Mexico must now stop at the border and transfer cargo to Mexican carriers, and Mexican carriers with cargo destined for points beyond the U.S. commercial zone must arrange for its transfer to U.S. carriers. Despite these restrictions, truck traffic between Mexico and the United States increased 42 percent from 1987 to 1990. In 1990, about 1.8 million commercial vehicles crossed the U.S.-Mexico border. Between 70 and 80 percent of all U.S. trade with Mexico is carried by truck.

NAFTA provides that the United States and Mexico will allow trucks access to each other's border states for the delivery and backhaul of cargo by January 1996. By 2000, all restrictions on cross-border trucking will be lifted. All cross-border charter and tour bus restrictions will be lifted in January 1994, when the Agreement goes into effect. By 1997, all restrictions on regular routes in scheduled operations will end.

Under NAFTA, Mexico will allow up to 49 percent Canadian and U.S. investment in carriers established in Mexico to transport international cargo and in bus companies by 1996 and up to 51 investment percent by 2001. By 2004, Mexico will end all restrictions on foreign investment in truck and bus companies except those related to truck service for the carriage of domestic cargo. Similarly, U.S. restrictions on Mexican-owned or controlled subsidiaries in the United States for transport of international cargo will end by 1998. Restrictions on Mexican-owned or controlled bus subsidiaries in the United States will be lifted in 2001.

NAFTA establishes a Land Transportation Standards Subcommittee to harmonize regulations relating to bus, rail, truck and hazardous material transport. This group will continue on a trilateral basis the work begun by the U.S.-Mexico Transportation Working Group.

Because NAFTA eliminates the need for cargo transfers at the border, a significant component of cross-border traffic congestion will be reduced. NAFTA-generated increases in U.S.-Mexico trade, however, will increase cross-border traffic. By 2000, upwards of 12 million trucks will cross the border each year, accompanied by corresponding increases in environmental problems stemming from emissions and noise.

Diesel fuel emissions include particulates and such gases as nitrogen oxides (NO_x), sulfur dioxide (SO_2), carbon monoxide (CO), and aldehydes. NO_x and SO_2 contribute to ozone formation and formaldehyde is a carcinogen. New engines will reduce particulate emissions by 83 percent by 1994 and NO_x emissions from 5.0 to 4.0 grams per brake-horsepower hour by 1998. In addition, SO_2 content of U.S. highway-use diesel fuel is limited to 0.05 percent by weight. Mexican truck engines based on the European 13-mode diesel emissions test offer a minimal level of emissions control. Moreover, Mexican diesel fuel contains significantly more SO_2 than U.S. diesel.

If efforts under the North American Automotive Standards Council, established by the NAFTA, are successful in making fuel standards compatible, Mexico may adopt U.S.-type standards and test procedures. This would mean that Mexican truck emissions would be the same as U.S. trucks and the only adverse air quality impact would come from increased traffic. However, if existing Mexican standards are not changed, Mexican vehicles operating in the United States would contribute more NO_x and particulate pollutants in the cities they transit. U.S. trucks operating with Mexican diesel fuel could experience a degradation of emissions performance, as high sulfur levels harm U.S. engine emissions treatment devices.

Cooperation among the NAFTA countries should eventually harmonize commercial vehicle emissions, engine manufacturing, and fuel standards, significantly reducing associated environmental problems.

Because Mexican trucks are not now required to meet U.S. engine manufacturing standards that reduce engine noise, larger numbers of Mexican trucks in the United States will increase associated noise problems. NAFTA efforts to harmonize safety and environmental standards are likely to lead to improvements in Mexican trucks that would reduce noise emissions.

Although the NAFTA includes a commitment from the United States, Mexico and Canada to work toward compatible technical and safety standards, it does not require the United States to change its size and weight limits, or indeed any of its regulations applicable to motor carrier operations. Any future changes in U.S. truck size and weight standards that might result from this process must be consistent with U.S. law. Current U.S. requirements governing truck size and weight are statutory and can only be changed by amending the relevant statutes. Moreover, truck safety standards will be enforced for Mexican trucks operating in the United States in the same way that they are enforced for U.S. trucks. Mexican drivers and vehicles will be required to meet all federal standards for driver qualification, training and licensing and for vehicle operations.

b. Rail transport

Rail shipment produces far less air pollution, is safer per ton-mile of freight carried, and achieves higher ton-miles per gallon than truck equipment used to carry the same commodities. The number of northbound railcars processed in the four U.S. Customs Districts on the Mexican border increased from approximately 71,000 in 1986 to 116,000 in 1990. NAFTA will likely result in increased use of rail movements to carry trucks between the United States and Mexico.

Although Mexico's Ferrocarriles Nacionales de Mexico ("FNM") is fully government-owned, U.S. railroads are frequently permitted to carry cargo (with their own equipment) directly to Mexican destinations. Plans are under way to expand the recently-introduced double-stack container service which has improved rail efficiency and reduced some truck traffic.

NAFTA ensures that U.S. railroads and intermodal companies will be able to continue to take advantage of gains made through informal agreements with Mexico to market services, operate unit trains, construct and own terminals, and finance rail infrastructure. Mexico continues to reserve ownership and operation of its rail system to its national railroad.

Most intermodal traffic moves by rail to the Mexican border and by truck within Mexico. The NAFTA should encourage continued FNM improvements to intermodal handling and terminals in Mexico. More business will be routed over the FNM, limiting road congestion. U.S. railroads and other private parties can participate in building intermodal facilities in Mexico, as long as the facilities are not on railroad property. Such investment should increase double-stack and other intermodal traffic, attracting more traffic away from motor carriers.

c. Hazardous material transport

Mexico is the third largest U.S. trading partner in chemicals, after Japan and Canada. In 1989, 6 percent of the U.S. chemical export trade, amounting to $2.2 billion, and almost

3 percent of the U.S. chemical import trade, amounting to $582 million, was with Mexico. The vast majority of hazardous materials are shipped by truck. All Mexican shipments of hazardous materials to the United States must be in full compliance with U.S. regulations.

Current restrictions in the movement of trucks across the border represent a significant environmental problem. The border transfer of hazardous materials from one truck to another is an especially dangerous practice and an unnecessary environmental risk. NAFTA provisions eliminating the need for border cargo transfers will significantly reduce this risk. In addition, harmonization provisions of the NAFTA specify that hazardous materials regulations are to be addressed by 2000, six years after NAFTA's implementation.

It is expected that the NAFTA could increase the flow of hazardous materials between the United States and Mexico, but it is difficult to quantify the extent to which such growth will increase safety and environmental risks. Certainly, the current safety record for hazardous materials transportation along the border is quite good. Since shipments must comply with U.S. hazardous materials transportation regulations, significant increases in accidents involving hazardous materials transportation are not expected. Mexican regulations are not presently as stringent as those in effect in the United States, but Mexico is planning to adopt regulations compatible with and equivalent to those in effect in the United States and is on a common timetable for adopting and implementing the U.N. international standards.

d. Border infrastructure

Currently, U.S. Customs is increasing manpower resources at the Mexican border to handle increased congestion and traffic flow problems. Most ports have added cargo hours of service and changed hours of service to handle peak seasons. Some local trucking arrangements now permit mutual access, but at great cost to efficiency. In Laredo/Nuevo Laredo, for example, trucks from each country may carry cargo across the border provided they return empty. There are also built-in delays that result from procedures that are required for trucks to be cleared for entry into the United States. The problems related to truck access restrictions are complicated by inspection agency staff limitations and a lack of coordination between U.S. and Mexican Customs at some ports.

Absent improvements in traffic flow, increased commercial traffic resulting from increased development could contribute to existing congestion problems along the U.S.-Mexico border. To keep pace with traffic increases resulting from increased trade with Mexico, the General Services Administration is working with the federal inspection agencies to complete a $360 million project for improvement and construction of facilities on the border, some associated with new bridges. Of this amount, almost $237 million will be spent at five ports that are expected to be most seriously affected by increased commercial traffic. The project will provide for new Customs import inspection lots and expanded primary inspection lanes at Brownsville, El Paso, Otay Mesa, Nogales, and Calexico. Completely new border stations are planned near Brownsville and Laredo in Texas; at Calexico, Andrade, and Tecate, in California; and at Naco, Arizona.

The FNM and the Southern Pacific are planning a new international railroad bridge crossing between El Paso and Juarez and a new integrated, multi-purpose, multi-user switching yard, 12 miles southeast of the Juarez in an industrial zone, to connect with the new bridge.

In addition to improved rail service and more efficient customs processing, the relocated combined facility should improve the environment in the centers of El Paso and Juarez.

Where such projects are federally funded, such as Customs facility improvements or expanded highway access, or are the product of international negotiations, such as international bridges or new ports of entry, federal environmental assessments are required, as well as state environmental impact assessments for highway and rail projects. Because of these safeguards, significant adverse environmental impacts can be avoided, and there may be some important environmental benefits resulting from construction of new or improved infrastructure projects.

Border congestion is an increasingly serious problem that can have a detrimental effect on environmental quality. Long lines of trucks waiting with engines idling to clear Customs on both sides of the border can contribute to air pollution problems at ports of entry. Two studies are being conducted to address existing and potential congestion problems under the Intermodal Surface Transportation Efficiency Act, which directs the Secretary of Transportation to conduct studies related to cross-border transportation with Mexico. Section 6015 requires that the Department identify existing and emerging trade corridors and transportation subsystems that facilitate trade among the United States, Mexico, and Canada. Section 1089 requires a study of the feasibility of an international border highway discretionary program to fund infrastructure improvements at border crossings. The studies are being directed by the Federal Highway Administration through a supplemental $2.4 million appropriation to the federal-aid highway program.

Taken together, the studies should result in recommendations designed to improve cross-border traffic flow, thereby reducing congestion and resulting air quality problems.

4. U.S. Environmental Technology and Services

With the passage of NAFTA, Mexican demand for environmental technologies and services will increase significantly, presenting substantial commercial opportunities for the U.S. environmental technologies industry and creating high-wage jobs for American workers. Implementation of NAFTA and the associated Environmental Agreement is expected to increase considerably environmental spending in Mexico -- from $2 billion in 1993 to well over $3 billion by 1997.

Currently, the United States exports $1 billion in environmental services and technologies to Mexico, accounting for approximately 60 percent of all Mexican imports in this field. This trade currently supports about 27,000 jobs in the United States.

Increased demand for environmental goods and services in Mexico will translate directly into rising numbers of high-wage jobs for American workers in the environmental sector. These jobs will be generated due to increases in total Mexican environmental expenditures, more open bilateral trade, a codified Mexican commitment to the environment, increased financial resources and U.S.-Mexican border cleanup initiatives.

By 1997, NAFTA is projected to double the number of U.S. jobs related to Mexico's growing environmental market, from over 27,000 based on current export figures, to 58,000 based on export projections of nearly $2.7 billion. By the year 2000, estimates are that NAFTA could increase U.S. exports of environmental goods and services to an estimated total of $3.8 billion. A total of 42,000 jobs are estimated to be generated nationwide due to NAFTA in activities related to those increased exports. The additional environmental exports will improve the earning potential of these U.S. workers since, on average, export-related jobs pay almost one-fifth more than other American jobs.

C. EFFECTS ON U.S. ENVIRONMENTAL LAWS

Much of the concern about NAFTA and the environment has focused on whether NAFTA will undermine the ability of the U.S. government and the states to establish and enforce their environmental, health, and safety laws. In fact, far from weakening environmental, health, and safety measures, the NAFTA and the supplemental agreements affirmatively encourage the three countries to improve standards and strengthen enforcement.

This section addresses NAFTA's effects on U.S. (both federal and state) environmental laws, including the likelihood of "upward harmonization" of those laws; how NAFTA handles challenges to U.S. environmental laws as trade barriers; questions of preemption; the relationship between NAFTA and international environmental agreements and U.S. laws implementing those agreements; NAFTA's treatment of process and production methods issues ("PPMs"); and NAFTA's effect on "green procurement" requirements.

1. Levels of Environmental Protection and Harmonization of Standards

As described more fully in Section II.A, NAFTA contains explicit provisions that protect the United States' right to determine those levels of environmental protection it considers appropriate, and ensure that harmonization of standards between the U.S., Canada, and Mexico will not lead to a lowering of those standards.

Some fears have been expressed that NAFTA will require the United States to lower its tougher standards to conform to international norms. To the contrary, the NAFTA specifically states that countries are free to use more stringent standards than the international standards. Moreover, both Chapters 7 and 9 also contain provisions recognizing each country's rights to set the level of protection of human, animal, or plant life or health it considers appropriate.

The Environmental Agreement provides further insurance against a weakening of U.S. standards, explicitly stating that each country is free to determine its own levels of environmental and labor protection for its citizens and committing the parties to work to improve their environmental laws. Article 3 of the Environmental Agreement commits each government to "ensure that its laws and regulations provide for high levels of environmental protection" and to "strive to continue to improve those laws and regulations." At the same time, Article 3 explicitly recognizes "the right of each Party to establish its own levels of domestic environmental protection and environmental development policies and priorities, and to adopt or modify accordingly its environmental laws and regulations."

In addition, the Commission for Environmental Cooperation has as a major component of its workplan the strengthening of "cooperation on the development and continuing improvement of environmental laws and regulations" through promoting information exchanges and developing recommendations on how to increase compatibility of standards "without reducing levels of environmental protection" (Env. Article 10.3).

2. NAFTA's Requirements for Sanitary and Phytosanitary Measures and Standards-Related Measures, and Their Relationship to Environmental Laws

Certain concerns have been raised often enough to warrant a detailed explanation of NAFTA's provisions designed to ensure that governments in enacting environmental, health, or safety standards do not create unfair barriers to trade, and why it would be difficult to challenge U.S. environmental measures using those provisions.

While granting the federal government and the states broad discretion to set their own environmental, health and safety standards, NAFTA does require governments to meet certain elementary requirements when applying laws and regulations to achieve the government's chosen levels of protection, in order to safeguard against trade protectionism disguised as a health regulation. As discussed in Section II.A, the NAFTA requires that sanitary or phytosanitary measures -- "SPS" measures, those measures related to agricultural pests and disease and contamination in food -- have a scientific basis and be based on a risk assessment appropriate to the circumstances (Article 712.3). It also requires that standards-related measures ("SRMs") treat the goods of another NAFTA party in a non-discriminatory manner (Article 904.3).

As an initial matter, it should be noted that very few U.S. environmental laws and regulations would ever meet the initial threshold criteria that could subject them to challenge under NAFTA. While some important environmental health and safety laws are SPS measures as defined in Chapter Seven, or standards-related measures as defined in Chapter Nine, most provisions of the key federal pollution control laws -- e.g., the Clean Air Act, the Clean Water Act, and the Resources Conservation and Recovery Act -- and their state-level counterparts are not. (Exceptions include the automobile emission control provisions of the Clean Air Act.) These pollution control laws are largely focused on controlling pollution from stationary facilities located within the United States. They do not set requirements with any direct bearing on trade in goods or services. Second, it should be noted that U.S. systems for SPS and standards-related measures are not protectionist and already conform to the NAFTA disciplines.

Challenges to sanitary and phytosanitary measures. There is concern that the scientific basis requirement for SPS measures will lead to challenges to U.S. environmental measures, and that challenges to such measures could prevail in a duel of competing scientific expertise. This is not the case. The term "scientific" is not separately defined in the NAFTA text. Accordingly, under general principles of international law, the term scientific is to be interpreted in good faith, using its ordinary meaning in context and in the light of the object and purpose of the NAFTA. Consequently, the ordinary dictionary meaning would apply.

It is clear that under the NAFTA, the requirement that measures be based on "scientific principles" and not be maintained "where there is no longer a scientific basis" do <u>not</u> enable dispute settlement panels to substitute their scientific judgment for that of the government maintaining the SPS measure. The question for the NAFTA panel is whether the government maintaining the SPS measure has "a scientific basis" for the measure. "Scientific basis" is defined in Article 724 as "<u>a</u> reason based on data or information derived using scientific methods."

The question will not be whether the measure is based on the "best" science or the "preponderance" of science or whether there was conflicting science. The question is only whether the government maintaining the measure has a scientific basis for it. The NAFTA SPS text recognizes that there is seldom, if ever, scientific certainty and consequently any scientific determination may require a judgment among differing scientific opinions. The NAFTA preserves the ability of governments to continue to make those judgments.

Concerns have also been raised that Article 712.3's requirement that sanitary and phytosanitary measures be based on "a risk assessment, as appropriate to the circumstances," opens environmental measures to challenge, including on grounds that state and local entities lack adequate resources to conduct risk assessments for all the measures they promulgate. Article 712.3 does not specify that the government adopting, maintaining, or applying the measure must conduct the risk assessment itself. For example, a state government could rely on a risk assessment conducted by the federal government, another state government, or a competent international organization.

In addition, the NAFTA requires each party to ensure that any SPS measure that it adopts is applied only to the extent necessary to achieve its appropriate level of protection, taking into account technical and economic feasibility. NAFTA's opponents have argued that the use of the term "necessary" in the text actually means "least trade restrictive." This is not true. The NAFTA's negotiators specifically discussed whether there should be a "least trade restrictive" test in the NAFTA, and all three countries agreed that this obligation would not be included. Rather, the NAFTA uses "necessary" to address how a health law or regulation that is in place is applied. It does not address the validity of the underlying health law or regulation, or the level of protection afforded by those laws. As is the case with "scientific," the term "necessary" is to be given its ordinary meaning in light of the context.

Challenges to standards-related measures. As with SPS measures, concerns have been raised that NAFTA will allow challenges to U.S. environmental standards-related measures, and that standards harmonization will result in lower levels of protection.

The provisions of Chapter Nine were specifically negotiated to be clear that there would be no "downward harmonization" of standards-related measures (that is, SRMs are not required to be "harmonized" down to the lowest common denominator). While governments are required to use international standards as "a basis" (but by no means necessarily the only basis) for their standards-related measures, Article 905.1 explicitly states that this is to be done only where the international standard would be an effective and appropriate means to fulfill the government's legitimate objectives. Article 905.3 also explicitly provides that nothing in this requirement "shall be construed to prevent a Party, in pursuing its legitimate objectives, from adopting, maintaining or applying any standards-related measure that results in a higher level of protection than would be achieved if the measure were based on the relevant international standard."

Article 906 on equivalence does not require that the United States change any particular standards-related measure. Instead, Chapter Nine creates a process by which the three NAFTA governments can try to reach greater compatibility of standards-related measures among their countries, but that does not require the United States to agree to any particular change in any U.S. standards-related measures. Moreover, Article 906.2 makes it explicit that the NAFTA countries' SRMs are to be made compatible "[w]ithout reducing the level of safety or of protection of human, animal or plant life or health, the environment or consumers."

Nothing in the NAFTA reduces safety standards in the United States or precludes the countries from issuing new regulations as needed. For example, Mexican and Canadian motor carriers operating in the United States must continue to conform to all federal and relevant state operating standards for vehicles and drivers.

Further protection for U.S. environmental measures is provided in Article 904. Under Article 904.4, no NAFTA country may adopt, maintain or apply a standards-related measure that would create an "unnecessary obstacle to trade between the Parties." Article 904 specifically provides, however, that measures whose demonstrable purpose is to achieve a legitimate objective, and that do not operate to exclude goods of another NAFTA country that meet that objective, do not create unnecessary obstacles to trade. "Legitimate objective[s]" are defined in Article 915 to include environmental protection and sustainable development. For example, non-discriminatory appliance energy efficiency requirements would have such legitimate objectives.

It also bears noting that in Article 903, the NAFTA countries affirm their existing rights and obligations under other international agreements, including environmental and conservation agreements, to which the NAFTA countries are party.

3. Relationship of NAFTA to State and Local Laws

From the beginning of the NAFTA negotiations, a fundamental objective of U.S. policy was to ensure that the NAFTA did not result in lowering U.S. health, safety,and environmental standards, including state and local measures. The NAFTA secured that objective.

The NAFTA does not impede the ability of states to provide environmental and health protection for their citizens. Indeed, the NAFTA will in no way diminish or impair the constitutional and legal rights of state and local governments to adopt, maintain, or apply measures to protect public health and the environment.

The NAFTA does generally apply to state measures, providing for state measures to enjoy the same rights and meet the same obligations as for federal government measures, such as non-discriminatory treatment of imported goods and publication of laws and regulations. The federal government in each of the three NAFTA countries is fully accountable for any state or provincial measures covered by the Agreement (Article 105). This provision is drawn virtually verbatim from the United States-Canada Free-Trade Agreement ("CFTA"), which has been in force since 1989.

Nothing in the NAFTA automatically preempts state law. For those few areas where the NAFTA negotiators considered that state measures might in fact be inconsistent with the NAFTA (investment and services provisions), the NAFTA provides a procedure for grandfathering such measures. That is, if the procedures are followed, those non-conforming state measures in the investment and services areas will be exempted from NAFTA's obligations.

State and local SPS measures and SRMs. Chapter Seven on SPS applies to state and local sanitary and phytosanitary measures in most respects. But it does so not by mandating compliance with federal law but by requiring that state and local sanitary and phytosanitary measures comply with the rules set out in the NAFTA. Just as the federal government will be free to maintain or change its laws, subject to NAFTA rules, so will state and local governments.

The NAFTA is drafted as a set of prohibitions. Unless the NAFTA prohibits a certain type of measure or practice, a NAFTA country is free to maintain or impose it. Since nothing in the NAFTA precludes states from maintaining or adopting sanitary and phytosanitary measures that are higher than federal sanitary and phytosanitary measures, they will continue to have the right to do so.

The NAFTA negotiators specifically used the plural "levels" in Article 712.2, which talks in terms of each NAFTA government's right to set the levels of protection that it sees fit, in part in order to account for the fact that each NAFTA country may have a multiplicity of levels due to differences among the states and between the states and federal government.

Challenges to state standards. It is the practice of the federal government to notify state governments directly and promptly upon learning that another government intends to challenge a state measure, including an environmental law, under the dispute settlement procedures of our international trade agreements. This practice will continue under the NAFTA.

The NAFTA dispute settlement procedures are the avenue for formal government challenges against a state measure. Those procedures are detailed in Chapter 20 of the NAFTA. These include consultations between the countries involved, recourse to the good offices of the NAFTA Trade Commission and, if no settlement can be reached, referral of the matter to non-binding arbitration.

In addition, there are a number of other procedures available to try to resolve disputes over either state or federal measures. For example, the NAFTA establishes trinational committees to facilitate consultations and avoid or settle disputes regarding standards. In the case of agricultural and food safety standards, the countries concerned may also have recourse to the good offices of relevant international organizations for purposes of resolving their differences.

As is the case under existing international trade agreements, USTR will work closely with state government representatives at all stages of a dispute settlement proceeding in preparing the defense of any state measure challenged under the NAFTA.

State and local compliance with NAFTA's obligations. Article 105 of the NAFTA requires each country to take "all necessary measures" to ensure that state and provincial governments observe those provisions of the agreement that apply to them. This language is drawn directly from the U.S.-Canada Free Trade Agreement, which has been in effect since 1989. It simply reflects an understanding among the three countries that they intend to apply the agreement equally at the federal and state level, with exceptions to that general rule in certain areas. This language is not meant to suggest -- and has not been interpreted under the CFTA to mean -- that state or provincial regulation must conform with federal standards.

Concerns have been raised that Article 105 somehow interferes with our states' ability to maintain measures to protect public health or the environment. Article 105, and any measures taken thereunder to secure observance by state and local governments of provisions of the NAFTA, will in no way diminish or impair the constitutional and legal rights of state and local governments to adopt, maintain, or apply measures to protect public health and the environment.

The implementation of Article 105, and the precise legal relationship between the NAFTA and a country's domestic law, is a matter for each participating government to decide. Nothing in the NAFTA requires the federal government to take legal action against state measures that NAFTA dispute settlement panels may determine to be inconsistent with trade obligations. Under the NAFTA, panel opinions are advisory only. If the defending country loses, it is not required to remove or change the offending measure. It may offer trade compensation instead or permit the other country to take retaliatory action of equivalent effect.

As has been the case under the GATT, in those rare instances where state rules may be successfully challenged under the NAFTA, the federal government will work cooperatively with the states to seek a satisfactory resolution of the matter. Under the NAFTA, each country will retain full discretion, under its own political and legal system, to determine how to satisfy its trade obligations.

Moreover, Article 105 does not apply to the NAFTA's Chapter 9 provisions on standards-related measures (Article 902.1). Instead, Article 901.2 provides that NAFTA governments will "seek, through appropriate measures, to ensure observance" of Chapter Nine's provisions by state and local governments.

For further discussion of federal-state issues and the NAFTA, see Chapters 1, 7, 9, and 22 of the Statement of Administrative Action.

4. Measures Implementing International Environmental Agreements

As discussed in Section II.A above, NAFTA has provisions explicitly giving the trade provisions in certain international environmental agreements precedence over conflicting NAFTA obligations, and provides a mechanism for adding other agreements to those listed in Article 104 and Annex 104.1. This procedure for amending Article 104 is an ongoing process, to respond to changes in international environmental and conservation law and to provide greater clarity of the NAFTA's relationship to other environmental and conservation agreements. To this end, the Administration has already obtained commitments from Mexico and Canada to modify Article 104, upon entry into force of the NAFTA, to include the Convention Between the United States of America and the United Mexican States for the Protection of Migratory Birds and Game Mammals, and the Convention on the Protection of Migratory Birds (U.S.-Canada). The Administration intends to pursue further discussions with Mexico and Canada to add other agreements to Annex 104.1.

There is some concern that international environmental agreements that are not listed, or U.S. measures implementing our obligations under such agreements, are threatened by NAFTA. Such is not likely to be the case. There is unlikely to be any conflict between these obligation sand the NAFTA. For instance, the NAFTA, in Article 2101, incorporates the general exceptions to the General Agreement on Tariffs and Trade ("GATT"). These exceptions generally provide for measures to protect human, animal or plant life or health as well as for measures relating to the conservation of exhaustible natural resources. The NAFTA countries agree, in Article 2101, that measures necessary to protect human, animal, or plant life or health include environmental measures and that exhaustible natural resources include both living and non-living resources. U.S. measures taken to implement international environmental agreements are likely to fall within these exceptions, whether or not listed in Article 104.

5. Processes and Production Methods

Another frequently expressed concern with NAFTA is that it threatens U.S. laws that restrict market access to products that are produced in a way that harms the environment or human health (process and production method measures, or "PPMs").

The NAFTA does not change U.S. ability to implement these laws. Rather, its implementation will facilitate resolution of these PPM issues. Consideration of PPMs is a high priority for the United States in carrying out the work plan for the Council under the North American Agreement on Environmental Cooperation. This involves very complex, and often sensitive, questions of how to address any environmental effects of products due to the processes or production methods associated with them. These questions include how the product was harvested, how it was processed, and what effects its production, consumption, storage or disposal will have on the environment.

These questions are of a global nature, not limited just to North America. Therefore, while the Administration is committed to taking them up with Canada and Mexico in the context of the NAFTA and the Environmental Agreement, a broader dialogue is also needed. Indeed, preparatory discussions are currently under way in the Organization for Economic Cooperation and Development ("OECD") to develop a sound analysis of PPMs, with active U.S. involvement in those discussions.

Another important step will be to engage the GATT, beginning with a post-Uruguay Round workprogram on the environment, which the Administration will be working to launch at the conclusion of the Uruguay Round. This work is expected to include a thorough examination of the adequacy of the GATT's substantive rules as they relate to PPMs. The United States' objective will be to ensure that countries are able to effectively address environmental objectives while not providing a means for arbitrary limits on trade.

In sum, while NAFTA and the side agreements do not solve the issue of PPMs, U.S. participation in these institutions will facilitate its resolution.

6. NAFTA and "Green" Procurement

Some have argued that the NAFTA constrains the ability of governments to pursue "green procurement" policies -- a reference, for example, to recycling provisions found in certain U.S. government procurement specifications. This is not the case as long as such provisions have not been established to discriminate against products imported from the other NAFTA parties. Furthermore, NAFTA's government procurement provisions in no way discourage procurement policies designed to encourage development of clean technologies or energy efficiency.

Chapter 10 deals with questions relating to government procurement. Article 1007 spells out the disciplines applicable to technical specifications for government procurement contracts. Technical specifications should not be used to create unnecessary obstacles to trade and should, where appropriate, be specified in terms of performance criteria rather than design or descriptive characteristics. Technical specifications are defined in Article 1025 to "mean specifications which lay down the goods' characteristics or their related process and production methods."

VI. SPECIFIC ENVIRONMENTAL EFFECTS

INTRODUCTION AND SUMMARY

Three agreements -- the NAFTA, the Environmental Agreement, and the Border Environmental Cooperation Agreement -- should be considered as part of an integrated approach to environmental issues if the NAFTA is approved. Section V reviews the potential consequences for the environment of the NAFTA and the two related agreements from a national, macro-economic perspective. This section addresses more specific concerns: What effects are the NAFTA and the side agreements likely to have in specific areas and on specific aspects of the environment? Since the U.S.-Mexico border region is an area where a substantial amount of trade-related economic activity already occurs and will continue to expand -- with or without the NAFTA - - and because many serious environmental problems are already occurring in this region, this section will tend to focus on environmental effects in the border region.

It should be emphasized that current conditions along the border result from the existing economic and political relationship between the United States and Mexico; they are not the result of NAFTA, which has yet to come into effect. The Clinton Administration and the administration of Mexican President Salinas acknowledge that serious environmental problems exist in both countries, and are committed to addressing those problems. The important question that remains is: Are prospects for environmental protection and improvement better with the NAFTA or without it? This section demonstrates that implementation of NAFTA will expand the public and private resources for pollution control and other environmental efforts and disperse industrial development away from the border region.

The border region. The land border between the United States and Mexico extends slightly over 2,000 miles, from the mouth of the Rio Grande near Brownsville, Texas and Matamoros, Tamaulipas on the Gulf of Mexico to San Diego, California and Tijuana, Baja California on the Pacific coast. The vast portion of the 250,000 square mile border area is a lightly populated desert or semi-desert. Its 10 million inhabitants -- 3 million in the United States, 7 million in Mexico -- live mainly in a number of "Sister Cities" across the border from each other.

Rapid population growth in the border area in the last 20 years has paralleled the expansion of the manufacturing and commercial base of the border cities in both countries, including the maquiladora plants in Mexico. With or without the NAFTA, population growth and economic development are expected to continue in the border region, and such growth and development will have environmental effects. Those effects deserve particular attention given the fragile, and in some places already stressed or degraded, nature of the ecosystems in the water-short desert areas.

Air quality. While air quality in much of the predominantly rural border region is good, the large Sister-City metropolitan areas suffer from poor air quality. The two largest U.S. border cities, San Diego and El Paso, do not yet attain all U.S. ambient air quality standards. Visibility is another air quality concern; U.S. National Park Service monitoring shows frequent degraded visibility in many park areas of the Southwest.

Whether or not NAFTA is approved, the upward trend in emissions will continue unless significant changes in socio-economic conditions or government pollution control programs occur. The sources of the upward tendency are easy to identify: factories will continue to be built on both sides of the border, and populations will continue to increase, adding more cars and homes. The socio-economic and legal corrective responses are less clear. In the absence of NAFTA, it is possible that the Mexican government would not meet the current and projected levels of SEDESOL investment in air pollution control in the border area and might devote fewer resources to cooperation with EPA to improve border air quality. If this were to happen, composite Mexican border emission increases could be in the significant to profound range.

With NAFTA implementation, on the other hand, it is reasonable to assume that Mexico will continue its aggressive border clean-up, as outlined in the U.S.-Mexico Border Plan and its successor, the Border Action Program. With an active Mexican air pollution control program, changes in emissions on the Mexican side of the border could range from a slight decrease to possible emissions increases that could be substantial at the highest industrial growth projections. Thus, without NAFTA Mexican border emissions affecting the U.S. may well be higher -- and perhaps substantially higher -- than with NAFTA.

Water quality and supply. Water issues are among the most critical environmental problem in the border area. Most of the larger communities along the U.S.- Mexico border obtain their drinking water from surface supplies, including the Rio Grande (Rio Bravo) and the Colorado River. Many plant and animal communities depend on fresh water in both the permanent rivers and the many intermittent streams.

Water supply presents a constant challenge in the arid and semi-arid conditions that prevail throughout the border area. Ample supplies of fresh water are essential for agriculture, for business and industry, for area residents, and for the plants and animals that inhabit water-dependent ecosystems in the region. An issue of critical importance in this regard is the need to assure adequate and safe supplies of drinking water to area residents, particularly in the unincorporated settlements that have sprung up on both sides of the border.

Pollution of surface water and groundwater is a significant environmental problem at this time in the border area. The sources of the pollution include untreated or inadequately treated domestic sewage and industrial wastewater, contamination from surface run-off, and contamination from mineral salts, fertilizers, and pesticides in irrigation return flows.

Without NAFTA, there could be some difficulty in sustaining, much less increasing, the current level of cooperation between the United States and Mexico in providing and regulating water quality-related facilities. Existing programs and some new initiatives for water pollution control in Mexico would presumably move forward even without NAFTA, but a substantial gap would remain between the projected needs and the financing that would be available without the NAFTA. A study for the International Boundary and Water Commission, for example, identified a dozen high-priority major wastewater treatment projects needed in the border area. With the NAFTA and the infusion of new financing support through the North American Development Bank that would be established, the prospects for designing and constructing these facilities quickly will be substantially improved.

Solid and hazardous waste. The effective management of hazardous waste generated in or disposed of in the border region has been a priority concern for the United States and Mexico in recent years. Under Annex III to the 1983 La Paz Agreement, most hazardous wastes generated by maquiladora facilities must be returned to the United States for disposal in a manner consistent with RCRA and other U.S. requirements. A Mexican presidential decree also prohibits the importation of hazardous wastes for disposal from the United States or other countries.

Data suggest that the current levels of hazardous waste entering the United States from maquiladoras in Mexico do not have a significant effect on capacity needs for hazardous waste management facilities in the border area or in the rest of the United States. Increased generation of wastes in either country could lead to increased capacity demand in the United States. Increased enforcement efforts by both countries with respect to maquiladoras may also lead to an increase in the return of hazardous wastes to the United States from the border facilities.

While the impacts of NAFTA on hazardous waste management do not appear to be significant, the impacts in the absence of NAFTA could well be substantial. Without the NAFTA, the trade benefits of the maquiladora program will continue to encourage location of facilities along the Mexican border. These facilities will continue to generate hazardous wastes that must be transported back to the United States for management. Thus, without NAFTA the United States will face an increasing demand for capacity to dispose of hazardous wastes generated in the maquiladora factories.

With the NAFTA, however, there will be a Border Action Program and a Border Environment Cooperation Agreement that will help develop and control waste management facilities in the border area. These programs would provide significant investment incentives for improvement of the environment in the border area and facilitate cooperative efforts to develop and implement an effective Mexican hazardous waste management program.

Public health and food safety. From the regulatory perspective of the Food and Drug Administration ("FDA"), the no-NAFTA scenario should have no major effect on current FDA operations to monitor the food imports along the border. In estimating the implications for health in border communities, however, the no-NAFTA projection is less optimistic.

Whether or not NAFTA is approved, it is presumed that additional development will continue to take place on the Mexican side of the border, with the concomitant increase in migration from the interior of Mexico to the border. Environmental pollution will continue to be concentrated in existing highly populated corridors and the stresses on the infrastructure and services on both sides of the border will continue. Despite the current level of effort by U.S. and Mexican authorities to provide more sanitation, potable water services and basic health care, a larger absolute number of people remain without adequate services. The key infectious diseases are expected to remain major problems at the same incidence levels; deteriorating sanitation and water supplies could result in cholera outbreaks in border communities which, thus far, have not occurred. Measles will probably continue to decrease as Mexico vigorously implements childhood immunization campaigns, but chronic diseases will increase slowly with the population increase and the absence of resources for effective prevention campaigns.

Without Mexico's commitment under NAFTA and the Environmental Agreement to greater emphasis on environmental enforcement, the environmental health issues could increase, particularly for victims of surface water and groundwater pollution. Health problems ranging from bacterial infections to toxic effects from hazardous waste could also increase. Health and environmental resources are currently inadequate on both sides of the border and can be expected to remain so without NAFTA.

Wildlife and endangered species. There are many areas on both sides of the international border where wildlife habitat (including wetlands) is under pressure as a result of increased industrialization, infrastructure development, and agricultural development. For instance, it is estimated that ninety-five percent of the lower Rio Grande valley's native habitat on the U.S. side of the border has been lost to agricultural and other development, and wildlife in the area is considered to be severely threatened. There are also significant adverse effects to endangered and threatened species in the border area due to activities affecting habitat on both sides of the U.S.-Mexico border.

The increase in trade and corresponding agricultural, industrial, and commercial development pressures as a result of NAFTA could result in a short-term increase in adverse impacts such as use and pollution of already stressed water resources, and further encroachment on wildlife habitat. Such stresses are likely to decrease in the long-term, however, due to joint cooperation and funding of actions as a result of NAFTA to resolve environmental problems. Moreover, if NAFTA results in geographical diversification of investment and industrial growth in Mexico, developmental impacts on wildlife habitat (including wetlands) and biodiversity should stabilize, thus providing an opportunity to improve existing conditions along the border to the benefit of both countries. Long-term positive benefits would be more dramatic on the Mexican side of the border than in the United States. The net long-term effect on wildlife habitat and biodiversity could be slightly positive.

NAFTA could also promote moderate positive, long-term benefits to threatened and endangered species on both sides of the border, through increased cooperation, joint efforts, environmental education, training, and public outreach. Many species could benefit from these efforts, including the Sonoran pronghorn antelope and desert tortoise.

If NAFTA is not implemented, the maquiladoras would continue to be concentrated in the border areas, thus exacerbating environmental conditions on the border, such as loss of habitat, adverse impacts to endangered and threatened species, and reductions in groundwater levels. Many of the increased opportunities for cooperation, training and enforcement of wildlife protection laws would be lost.

The border area has been and remains a growing region of legal and illegal commercial plant and wildlife traffic for which detection and apprehension of illicit commerce is increasingly difficult. The increased flow of goods between the United States and Mexico under NAFTA will require increased law enforcement capability to address an expected increase in legal and illegal wildlife importation and commerce of wildlife products.

Actions to address the problem of increased trade are under way. In 1991, Mexico became a member nation of CITES, through which the two countries have begun to work more closely to increase wildlife enforcement along the border. Under the Environmental Agreement, all three countries are encouraged to work more closely on law enforcement, training, and educational projects. The net long-term effect should be reduced illegal trade, and thus, slight to moderate increased protection of endangered and threatened species.

Fisheries. All three NAFTA countries manage marine fisheries within exclusive economic zones 200 miles from their coasts. Since obligations derived from fishery management laws are excluded from the dispute resolution provisions of the NAFTA Environmental Agreement, challenges to the fishery management regime of each country will not occur under the NAFTA.

Some fishery resources span or migrate across borders. While cooperation in the management of fisheries has been steadily increasing, particularly with Canada, the commitment to cooperation and mutual economic benefit in NAFTA supports an acceleration of coordination and exchange of data and information useful to each country's fishery management system. The United States believes in managing fisheries throughout their range. The enhanced cooperative atmosphere under NAFTA and the Environmental Agreement will facilitate that approach and the quality of fishery management decision making affecting stocks of mutual interest.

Commercial exploitation of fishery resources will continue to be by national fleets. NAFTA will not lead to an influx of commercial fishing vessels from one country fishing in another's economic zone. The main source of new fishing pressure may come from increased tourism and associated recreational fishing.

The performance of the Mexican tuna fleet in reducing dolphin mortalities has improved dramatically because Mexico has domestically implemented dolphin conservation measures for its fleet. Mexico's progress with respect to conservation of sea turtles has also been noteworthy. Mexico has since 1991 implemented and enforced enhanced provisions to protect sea turtles caught incidentally in its shrimp-trawl fishery.

NAFTA has provided the incentive for many significant changes in Mexico's conservation policies, which will be further reinforced by the Environmental Agreement's cooperative activities. Without NAFTA, current bilateral consultations will continue, but without the improved atmosphere and high-level attention that the Environmental Agreement provides.

Forests, parks and rangelands. Mexican forest resources are an important national and global resource. Mexico ranks seventh in tropical forest area. Deforestation in Mexico is therefore a major environmental concern. Although the agricultural land use patterns and economic conditions that underlie most deforestation will likely continue regardless of the NAFTA, NAFTA could slow the rate of deforestation somewhat by facilitating a general increase in wage rates and an increase in alternative sources of productive employment. If the NAFTA is not implemented, it is unclear what alternate mechanisms to reduce deforestation could be applied.

The U.S. National Park System already faces pressures on its resources in the border area from increases in cross border transportation, increased population settlement with some additional industrial site development at or near the border, consequent decreased water quality, and further loss of cultural resources. Similar pressures confront Mexican park areas and their newly-created biological reserves.

Developmental pressures are expected to continue with or without the NAFTA, but direct public use and demands on public lands are likely to be at somewhat lower levels without the NAFTA. With the NAFTA, however, the level of long-term adverse impacts to park resources may be slightly reduced due to greater joint coordination and cooperation between U.S. and Mexican environmental agencies.

Only minor impacts are expected to public rangelands along the U.S. border with implementation of NAFTA. Other public lands along the border are already receiving ever-increasing amounts of recreational use. Generally, recreational users of public land are providing a much needed economic boost to the small communities on both sides of the border. NAFTA will cause little change in these trends.

Conclusion. The current environmental situation and trends for the future pose a serious challenge to the governments and citizens of both Mexico and the United States. They clearly call for the commitment of new resources to correct current conditions and properly manage future growth and development. While estimates of the cost to meet environmental needs in the border area vary widely according to the breadth of the programs envisioned, a figure of approximately $8 billion over the next ten years provides a reasonable benchmark.

New resources alone will not protect the environment. Private and public programs and institutions to channel and supplement those funds, to provide legal and administrative support, and to address changing circumstances are also essential to the task. Some programs and institutions are already in place, such as the cooperation between the two federal governments through such mechanisms as the La Paz agreement and the International Boundary and Water Commission. Clearly, however, a broader and more intensive focus on environmental problems calls for new initiatives and new institutions. The NAFTA, the Environmental Agreement, and the Border Environmental Cooperation Agreement are the kinds of initiatives required.

A. PRODUCT STANDARDS, PESTICIDES, AND FOOD SAFETY

NAFTA's provisions ensure that the ability of the United States to establish and enforce its food safety and pesticide standards will be maintained and that the integrity of U.S. regulatory processes will be fully respected. Through the NAFTA and the Environmental Agreement, enhancement of product standards and enforcement activities will be promoted in all three countries. NAFTA also provides strong incentives and an excellent opportunity to share expertise and experience for real public health and environmental gains.

1. Results of 1992 Recommendations

The 1992 Environmental Review recommended the following general goals for U.S. negotiators:

- "The U.S. will maintain the right to exclude any products that do not meet its health and safety requirements and the U.S. will continue to enforce these requirements;" and

- "The U.S. will maintain its right to impose stringent pesticide, energy conservation, toxic waste and health and safety standards."

In addition, the Review made several more specific recommendations to:

- maintain the right of each party to set standards;

- protect existing environmental and public health legislation and international agreements;

- protect against imports of goods that do not meet U.S. standards;

- improve cooperation between the United States, Mexico and Canada on health and environmental standards and pesticide regulation and management;

- agree that efforts toward harmonization should include the presumption that there be no diminution in protection of public health and the environment;

- provide for open dispute settlements involving participation by scientific experts; and

- agree that risk assessment and risk management be based on sound science.

Throughout the NAFTA negotiations the United States observed the major principle that the resulting agreement must fully preserve the integrity of U.S. regulatory systems for public health and environmental protection.

2. Background

As described in Section II.A, sanitary and phytosanitary ("SPS") measures are rules and standards that countries establish to protect human, animal, or plant life or health from pests, diseases, and risks posed by additives or contaminants. Pesticide residue tolerances -- the standards that EPA sets for maximum permissible residue levels in food -- are a prime example of an SPS measure. Chapter Seven of the NAFTA covers SPS measures. The NAFTA provisions will not require any changes to the U.S. regulatory framework. The United States operates an open, transparent, and non-discriminatory system. Under the NAFTA, the United States will continue to base its regulatory decisions on sound science, to establish pesticide residue limits and other regulatory requirements in a manner protective of the U.S. public and the environment, and to inspect at the borders for compliance with U.S. standards. If a U.S. standard is challenged, the burden of proof will be on the challenging country.

3. Potential Impacts of the NAFTA

There is concern that the NAFTA might require the United States to accept food imports that do not meet U.S. standards, or that U.S. standards might be successfully challenged under the NAFTA. Another fear is that efforts to harmonize these health and safety standards under the NAFTA might require the United States to lower its existing standards.

a. Pesticides

In the United States, pesticides are subject to a comprehensive regulatory scheme administered primarily by the EPA. EPA is responsible for registering all pesticides under the Federal Insecticide, Fungicide, and Rodenticide Act ("FIFRA"). The registration process results in a set of very specific terms and conditions under which a pesticide may be used. No pesticide may be imported into the United States for use in this country unless it is registered under FIFRA. Under the Federal Food, Drug and Cosmetic Act ("FFDCA"), EPA also sets tolerances, or maximum permissible residue levels, for all pesticides used on food or feed crops. These tolerance regulations apply equally to all domestically-produced and imported foods. Tolerances are monitored and enforced for both domestically-produced food and food imported into the United States by the U.S. Department of Agriculture ("USDA") and the Food and Drug Administration ("FDA")

Since both registration and tolerance-setting are based on an evaluation of the safety of the pesticide, the agency generally coordinates its review and approves registrations for specific crop uses along with tolerances for specific foods. EPA registers pesticides and establishes tolerances on the basis of data contained in registration applications and tolerance petitions. The agency has detailed regulations spelling out the types of studies required and appropriate test methods. EPA has also implemented policies to assure data quality, through Good Laboratory Practices ("GLP") regulations. The regulations apply equally to domestic and foreign laboratories that develop data in support of U.S. registration. The data submitted must establish that the pesticide will not cause unreasonable adverse effects on human health or the environment.

Although no pesticide may be used on a food crop or processed food in the United States unless it is registered for that food, it is possible to establish a tolerance without a corresponding registration. This could be the case, for example, if a crop is not widely grown in the United States or a pesticide is used abroad to deal with a pest problem that may not be present here. Such tolerances are often referred to as "import only" tolerances. So long as there is a U.S. tolerance, food containing residues within the tolerance limitation may be legally imported into the United States, even if there is no registration for use in the United States. In the absence of a U.S. tolerance, however, foods with residue of a pesticide may not enter the United States, even though there may be a tolerance or registration in the country where the food is produced.

b. Meat and poultry

The Food Safety and Inspection Service ("FSIS") in the Department of Agriculture ensures that domestic and imported meat and poultry products are safe, wholesome, and accurately labeled. This responsibility derives from the Federal Meat Inspection Act ("FMIA") and the Poultry Products Inspection Act ("PPIA"). There are no provisions in NAFTA that would compromise the level of protection to human health now provided for under FMIA and PPIA.

All imported products are subject to requirements that are at least equal to those applied to products produced in the United States. NAFTA recognizes, however, that there may be different, but equivalent, scientifically justifiable methods of achieving sanitary requirements. This does not represent any concession of requirements. Rather, the NAFTA permits each country to retain its own appropriate level of protection with respect to imported products.

Mexico is the second largest importer of U.S. meat and the third largest importer of U.S. poultry. Current imports from Mexico account for less than one percent of all U.S. meat and poultry imports. The volume of future exports from Mexico to the United States will depend greatly on demand in Mexico relatively independent of NAFTA, depending on the economic growth and an expanding Mexican population. It is likely that these domestic factors will moderate any meat and poultry exports from Mexico and that FSIS current resources will be able to address any future patterns of imports of these products from Mexico.

Canada is the third largest market for U.S. meat exports and the fourth largest market for U.S. poultry exports. In addition, Canada is also one of the largest exporters of meat and poultry products to the United States.

A continuing trend toward harmonization of reinspection systems and many other positive regulatory and trade effects have ensued since the implementation of the U.S.-Canada Free Trade Agreement ("CFTA"). A U.S.-Canada Technical Working Group has been established to examine the whole range of import activities in both countries in order to improve procedures currently in place. The Working Group is also an effective instrument for dispute resolution. The health, safety and quality standards of FSIS were not compromised under the CFTA. FSIS believes that these standards will not be compromised under NAFTA.

c. Other products

Under the authority of the FFDCA, FDA's activities are intended to assure consumers that drugs and medical devices are safe and effective for their intended uses; that cosmetics are safe and made from appropriate ingredients; that all labeling and packaging is truthful and informative; and that foods are safe and wholesome and produced under sanitary conditions. FDA formulates and enforces regulations and guidelines for product standards and specifications, such as those covering chemical and microbiological contaminants, good manufacturing practices, and labeling requirements. FDA's laws and regulations are applied equally to domestic and imported products. Products that do not meet those laws and regulations may not enter the United States. Because the provisions of NAFTA ensure that each country will continue to be able to apply its health measures in a manner that fully achieves its chosen level of health protection, FDA will continue its border enforcement activities.

FDA records indicate that the violation rate for imported Mexican produce with illegal pesticide residues, 3.8 percent, is comparable to that of produce from other countries. Most of these residue violations involve residues of pesticides that are registered in the United States, but not for use on the food crop in which the residue was detected.

The Mexican government works with FDA and EPA as a partner in efforts to reduce the likelihood that violative residues will appear on food imported from Mexico. The Mexican government and Mexico's agricultural and pesticide industries have worked with the United States to exchange information on FDA's regulatory system, the results of FDA's import monitoring, and on how to comply with U.S. pesticide tolerances. Mexico's Ministry of Health has an annual bilateral meeting with FDA, and Mexico and the United States have developed a jointly agreed workplan for technical cooperation on a variety of issues and other border conferences on matters of mutual interest, as well as telephone communications on a daily basis. These cooperative activities demonstrate Mexico's commitment to assuring that its pesticide uses are made consistent with U.S. requirements.

B. AIR QUALITY

With or without the NAFTA, increasing industrialization and population growth is likely to lead to increased air emissions in the U.S.-Mexico border area.

It is estimated that, within eight to ten years of implementation of the NAFTA, border area air emissions in Mexico are likely to be less than emissions levels in the absence of NAFTA. Moreover, implementation of the NAFTA will provide added impetus to cooperative programs already underway between the U.S. and Mexican governments to improve air quality in the border region.

Without NAFTA, the continuation of the maquiladora program is likely to result in more rapid growth in air emissions in the border region, since incentives will be maintained for companies to build additional facilities near the U.S.-Mexico border. Such facilities are likely to be associated with further population increases, which will, in turn, contribute additional sources of air pollution in the form of motor vehicle emissions and pollution from residential heating fuels.

1. Results of 1992 Recommendations

The 1992 Environmental Review discussed a number of options for addressing air pollution in the U.S.-Mexico border region. Many of these activities are being undertaken pursuant to the Integrated Environmental Plan for the Mexican-U.S. Border Area, which was released in February 1992. Current activities include:

- the establishment of air pollution monitoring programs in Ciudad Juarez-El Paso, Mexicali-Imperial Valley, Tijuana-San Diego, and Nogales;
- the provision of air quality training to Mexican officials; and
- the development of emissions inventories and monitoring networks for priority binational air basins, in order to determine cost-effective air pollution control strategies, and to measure progress and compliance.

These programs are described in more detail in Section IV.A.

The 1992 Environmental Review included an extensive discussion of Clean Air Act measures being carried out in the U.S. area of the border region. These measures continue to be implemented, and are expected to result in an overall decrease of 20 to 30 percent in air emissions on the U.S. side of the border over the next ten years.

2. Background

Air quality in most of the predominantly rural border region appears to be quite good. In the larger Sister-City metropolitan areas, however, poor air quality is a reality. The two largest U.S. border cities, San Diego and El Paso, do not yet meet all U.S. ambient air quality standards. Due to dust entrainment and some industrial effluence (which often originates on the Mexican side of the border), portions of six other border counties in California, Arizona and New Mexico (which are largely rural) also fail to meet all U.S. air quality standards.

Visibility, an air quality value not strictly related to human health, is another concern in the border region. The U.S. National Park Service's visibility monitoring network shows a sizable number of days with degraded visibility in many park areas of the Southwest. Some (although not all) of the sources of these visibility problems are located on the Mexican side of the border.

a. Carbon I and II

The United States has recently become concerned about high levels of sulfur dioxide emissions from Mexico's Carbon I and II power stations, which are located 20 miles south of the Texas border. Emissions from these stations may have a significant adverse impact on visibility in Big Bend National Park in the United States, 130 miles away.

Both plants were sited and permitted in accordance with Mexican law. Moreover, construction on both plants commenced prior to the NAFTA negotiations, and demand for the power they generate will exist in Mexico regardless of whether or not NAFTA is implemented.

Over the past six months, U.S. officials have raised their concerns with Mexican officials at high levels. On October 26, 1993, in the context of a Ministerial meeting pursuant to the Integrated Border Environmental Plan, the United States and Mexico agreed to establish a binational technical work group. This group has been directed to develop bilateral measures to preserve air quality and to address existing situations of substantial air quality degradation, including visibility problems at Big Bend National Park.

The work group has been directed to analyze air quality and visibility problems related to the Carbon II plant, as well as to other sources, and to make recommendations for an equitable reduction strategy to address these problems. The work group will issue its initial report in January 1994.

b. Other industrial sources

Sulfur dioxide (SO_2) emissions from copper smelters and utilities on both sides of the border have been a concern in the past. However, the situation has been improved as a result of cooperative efforts between the U.S. and Mexican governments under Annex IV, adopted in 1987, to the 1983 U.S.-Mexico Border Environmental Agreement. This agreement resulted in a standard emission limitation on U.S. and Mexican border copper smelters.

The United States and Mexico are currently compiling emissions inventories for maquiladoras. Officials are particularly concerned about emissions of volatile organic compounds ("VOCs") from plants that manufacture electronic and electric equipment, material, and supplies, transportation equipment, and furniture. VOCs are major precursors of ozone formation, and they contribute significantly to air toxicity. Other types of industries located in the border area which contribute to high levels of air pollution include oil and gas, metallurgy, iron and steel, electric power generation, cement manufacturing, mining, and brick manufacturing.

c. Motor vehicle emissions

Motor vehicle emissions are another major source of air pollution. Results from a joint SEDESOL-EPA vehicular testing program conducted in Juarez in late 1990 showed the Juarez fleet to suffer from a very high level of tampering with emissions control equipment, which significantly reduces combustion efficiency. Inefficient automobile fuel combustion results in high emissions of VOC, nitrogen oxides, carbon monoxide, and particulate matter. It is likely that similar problems exist in other border towns.

d. Residential sources

Many residences in the Mexican border area burn non-traditional fuels to provide warmth in the winter. Such fuels include wood scraps, cardboard, and tires. In adjoining U.S. areas in the same airshed, constricted airflow due to terrain and/or temperature inversions exacerbates the effects of these emissions, resulting in dangerously high levels of particulate matter and carbon monoxide. This problem is particularly marked in the El Paso-Juarez airshed.

3. Potential Impacts of the NAFTA

a. General border area effects

It is likely that industrial growth will continue in the border area with or without a NAFTA, and that there will be associated air impacts along the border. The 1992 Environmental Review included detailed projections for growth in the border area with and without NAFTA. The following discussion summarizes these projections.

All forecasts of border area economic growth have a substantial range of uncertainty. Based on the volatility of past growth patterns and ongoing changes in Mexico's economy, plausible baseline forecasts of border growth, therefore, could range between 5 and 15 percent per year absent NAFTA. With the removal of trade barriers that would accompany a NAFTA, the already brisk U.S.-Mexico trade will be enhanced. However, with NAFTA the relative economic advantage of the Mexican border region over cities in the Mexican interior is likely to be considerably less than it is under the current maquiladora program.

Therefore, it is possible to specify two plausible scenarios to express the effect of NAFTA on the baseline economic projection.

Under **Scenario 1**, the relative weight of the border in Mexican growth is unchanged. Under **Scenario 2**, NAFTA leads to relatively more investment in the Mexican interior. The former projection might reasonably result in a 6 to 17 percent annual growth rate in Mexican border industries, while the latter projection might be expected to result in a 4 to 12 percent annual industrial growth rate in the Mexican border region.

Under either scenario, the number of Mexican pollutant-emitting facilities will increase, although at different rates. Border growth will also prompt increases in Mexican commercial and residential pollutant sources, and in U.S. Sister Cities as well.

Another factor which must be considered is the degree of policy cooperation between the United States and Mexico in controlling border area air pollution. Different levels of control would, of course, have an important impact on air quality, in addition to the growth in the number of sources.

Scenario A would result if the growth-related pollutant increases in Mexico are unabated due either to failure of the new, proposed SEDESOL control initiatives or to failure to implement these initiatives. **Scenario B** would result if the currently planned EPA-SEDESOL air technical initiatives enunciated in the Border Plan are carried out expeditiously.

Scenario B is considered likely if the U.S. and Mexican governments continue to follow up quickly on the commitments contained in the Border Plan, as is expected. Actions which could be expected to occur under Scenario B include: enhanced industrial inspections and enforcement; tighter emissions standards for new automobile emissions; start-up of vehicular inspection and maintenance ("I/M") programs in Juarez and Tijuana; and retrofitting and retirement of existing, highly polluting industries (such as brick manufacturing facilities in Juarez).

Under Scenario B, an initial, short-term emissions increase (which would probably result in some worsening of U.S. border air quality for at least some locations), would likely be followed by a rapid reduction in air pollution concentrations.

Scenario A is most likely to occur in the absence of NAFTA, since the Mexican government would have less financial support for air quality programs, and limited Mexican revenues would constrain available resources for air emissions control.

In any projection, growth in the immediate U.S. border area should also be expected. This growth should be less pronounced than in Mexico but is likely to be significant, particularly for existing large and mid-sized border communities. However, the more stringent U.S. air quality requirements that are already being implemented are likely to cause air emissions to continue their long-term decrease through the 1990s.

Industrial emissions and residential emissions in the United States may increase slightly, but it is likely that these will be offset by continued decreases in vehicular fleet emissions. Thus, in all three projections described below, air emissions from U.S. border areas are expected to experience an overall decrease of 20 to 30 percent over the next ten years (a decline approximately equal to historical trends in the 1980s).

In the following paragraphs, the baseline growth estimate and two different NAFTA border area growth scenarios are assessed for their implications for air emissions changes under varying policy cooperation scenarios. These air emissions estimates are extrapolations given historical trends and the current state of knowledge about Mexican air emissions. They should be interpreted as illustrations, not forecasts. The composite air emission estimates are based upon an overall assumption that 20 percent is industrial, 50 percent is vehicular, and 30 percent is residential/ commercial. Table 2 summarizes these estimates.

b. Border effects with NAFTA: Scenario 1 (current proportion of Mexican growth continuing in the border area)

With expected annual Mexico border industrial growth of between 6 and 17 percent, over ten years there would be a cumulative increase in Mexican border industry of between 79 and 380 percent. If past population-maquiladora relationships hold, the Mexican border population would increase by approximately 50 to 230 percent in ten years. With such growth adjacent to the U.S. border, it is highly likely that Mexico would concentrate abatement efforts along the border; therefore, policy cooperation Scenario B is much more likely. A Scenario B future could reasonably hold such industrial emission increases to 40 to 250 percent over those ten years, depending upon the magnitude of the industrial growth.

Even with a significant population increase and increases in vehicle miles traveled, Mexico's requirement for catalytic converters on Mexican-produced automobiles, coupled with aggressive implementation of vehicle I/M programs, could result in a sizable decrease in vehicle pollution (30 percent) under the lower range growth rates. However, with growth at the higher end of the plausible range, there would still be a significant increase in vehicular air emissions (perhaps as high as 120 percent).

For residential and commercial emissions, the growth projections under Scenario 1 could translate into an approximate 20 to 180 percent increase in such emissions if control Scenario B is a reality. Such an increase could occur despite Mexican government subsidies for cleaner heating fuels and upgraded residential housing.

Thus, under the growth Scenario 1 projection, in ten years the composite Mexican border air emissions impacting the United States in the most likely control scenario (Scenario B) might vary from little or no change in the case of the lower industrial growth rates to profound change (over 150 percent) in the case of the higher industrial growth rates.

c. Border effects with NAFTA: Scenario 2 (proportionally more growth in the Mexican interior than in the border area)

With expected annual industrial growth in the border area of 4 to 13 percent, there may be a cumulative increase of between 48 percent and 239 percent in Mexican economic activity in the border region within ten years. If past maquiladora industrialization-population trends hold, a Mexican border population increase in the range of 30 percent to 150 percent could be expected.

With aggressive inspections and enforcement (under Scenario B), industrial emissions might still rise 20 percent to 150 percent in ten years. With increasing integration of the border economies, Mexican industrial fuels will probably increasingly tend toward natural gas, as is common on the U.S. side of the border. (As discussed in Chapter V, NAFTA's provisions

TABLE 2

PROJECTED CHANGES IN BORDER AIR EMISSIONS 1994 - 2004 UNDER THREE GROWTH PROJECTIONS (ALL ESTIMATES ARE APPROXIMATE)

Projection	Mexican Composite Air Emissions Change
With NAFTA --	
Growth Scenario 1 and Policy Cooperation Scenario B	No change to profound increase (0% - 165%)
Growth Scenario 2 and Policy Cooperation Scenario B	Moderate decrease to significant increase (-20% - +85%)
No-NAFTA	
-- Policy Cooperation Scenario A*	Significant to profound increase (40% - 225%)
-- Policy Cooperation Scenario B	Slight decrease to profound increase (-10% - +125%)

Under all three growth projections, border air emissions from U.S. sources would experience slight to moderate decreases (20%-30%) between 1994 and 2004.

* Scenario A denotes a reduced level of Mexican air pollution control; Scenario B denotes a high level of Mexican initiative; see text.

may encourage such a shift.) Greater use of natural gas would significantly reduce CO_2 emissions (an important global-warming gas) over those experienced with fuel oil or coal.

The most significant advantages from enhanced SEDESOL activity should come from controls on vehicles. These controls include: a requirement for catalytic converters; tough new vehicle emissions standards that correspond to U.S. levels; an inspection/maintenance program; and increased availability and use of unleaded gasolines in Mexico within the next few years. Under Scenario B, vigorous enforcement of these measures could reasonably be expected to produce a change in the Mexican border fleet emissions of between -50 percent and +50 percent in ten years, depending upon the degree of industrial growth.

The likely change in emissions due to residential/commercial growth is less clear. With increasing employment and personal wealth, Mexican citizens should be able to attain higher living standards that would decrease the need to burn non-traditional fuels in residences. In this context, implementation of the NAFTA may be critical. The additional economic growth arising from economic and trade opportunities under NAFTA may allow job creation to match or exceed labor force growth; this would allow for the income increases that would stimulate Mexican consumers to switch from higher-polluting, non-traditional fuels.

Given a reasonable schedule for construction of new, basic housing, residential/commercial emissions under Scenario B could increase 10 to 100 percent within ten years.

Thus, in ten years, under the most likely air emissions control scenario (Scenario B), changes in composite Mexican border air emissions that could affect the United States may range between an approximate 20 percent decrease to an 85 percent increase.

It should be noted that Mexican emissions rates are likely to change over time. Specifically, it is likely that air emissions would continue to increase following implementation of the NAFTA, and that enhanced air quality control efforts will result in gradual reductions in these emissions as the 1990s progress.

d. Effects on the U.S. interior

NAFTA could cause a variety of transportation-related air quality impacts on areas of the United States beyond the immediate border region. These could be caused by:

- increases in U.S. registered motor vehicles (automobiles and trucks) entering Mexico and returning to the United States;

- increases in Mexican registered motor vehicles entering the United States and returning to Mexico; and

- increases in U.S., Mexican and Canadian motor vehicles transiting the United States between Mexico and Canada.

Mexican automobile emissions have been poorly controlled for many years. However, Mexico is now enforcing emissions standards for both domestically-built and imported vehicles. In addition, SEDESOL has promulgated emissions standards for 1993 that correspond to the current U.S. standards. SEDESOL plans to further reduce its emissions standards in 1994-96. The gasoline most commonly available in Mexico is leaded, although unleaded is available in larger cities and on main trunk lines. In December 1991, the Mexican government cut the price margin between leaded and unleaded fuels in half. In addition, progress has been made in increasing the availability of unleaded fuel.

Mexico's truck fleet is generally powered by diesel engines, with no significant level of emission control. Mexico is currently considering establishing an emissions standard for trucks based on U.S. standards, which would lead to better emission control. The diesel fuel found in Mexico is generally of a good octane level but very high in sulfur. There is reportedly a high level of particulate contamination in Mexican diesel fuel. Mexico has plans to reduce the sulfur level of at least a part of its diesel fuel pool.

It is expected that the NAFTA will encourage a continued high level of cooperation on pollution control between the United States and Mexico. Therefore, this analysis assumes that Mexico will adopt: (1) more stringent standards comparable to U.S. standards, which are due to take full effect in the United States in 1996; (2) better fuel quality regulations; and (3) better I/M programs. Furthermore, it is expected that more new catalyst-equipped vehicles will be purchased by Mexican citizens, and that these vehicles will be maintained better, possibly encouraged by a better I/M program in Mexico. Therefore, while increased prosperity is likely to result in more trips by Mexican citizens to U.S. cities away from the border, this increase in vehicle miles traveled ("VMT") is not likely to result in a major increase in overall emissions.

While it is difficult to predict with certainty the impact of NAFTA on the Mexican automotive fleet and fuels, it is certain that a well-controlled (and maintained) fleet and good quality fuels cost more money than poor quality fuels and vehicles with catalytic converters. If NAFTA results in greater prosperity for Mexico, then this should result in more money being available for well controlled and maintained vehicles and better fuels. This will, in turn, benefit both countries.

Another area of concern is the improper use of Mexican leaded fuel by U.S. light-duty vehicles which occasionally transit Mexico, with resulting damage to catalytic control systems. It is expected that this will not be a significant problem in view of the increased availability of unleaded gasoline in Mexico discussed above.

As described in Section V.B.3, U.S. trucks are not permitted over the border into Mexico, and Mexican trucks are permitted in the United States only as far as the border commercial zone. NAFTA will allow trucks to transport cargos directly to their destinations in both countries. In assessing the impact of this change, it is assumed that the NAFTA will result in an equalization of heavy-duty ("HD") engine emission standards between Mexico and the United States. It is also assumed that Mexico will implement a low-sulfur diesel fuel requirement, due to the negative impact of fuel sulfur on expected future HD diesel aftertreatment devices. (This is a reasonable assumption, since NAFTA establishes a North

American Automotive Standards Council, which will develop a work program to make fuel standards compatible.)

With U.S.-type standards (and test procedures) and low-sulfur fuel, Mexican trucks operating in the United States would emit pollutants at roughly the same rate as U.S. trucks. The only negative impact on air quality would result from any net increase in overall truck traffic resulting from increased economic activity. If, however, Mexico does not adopt U.S.-type standards and test procedures, then Mexican vehicles operating in the United States would contribute more NO_x and particulates, and U.S. trucks could experience degradation in emissions controls as a result of operating with Mexican diesel fuel with a high level of sulfur.

4. Likely Trends in the Absence of NAFTA

a. Effects on the border

Without NAFTA, it is expected that the maquiladora program will continue to provide incentives for the construction of facilities in the border region, and that Mexican border populations will continue to increase, with resulting increases in air pollution emissions. Wages would be less likely to improve in a relative sense, so that widespread retirement of highly polluting automobiles and residential fuels would be less likely to occur. It is not clear, however, whether efforts by the Mexican government to control emissions will continue to increase without NAFTA. There are two possible outcomes.

First, due to budgetary constraints, the Mexican government may be unable to maintain current levels of cooperation between SEDESOL and EPA aimed at improving border air quality. Moreover, the Mexican government may choose to invest its resources in Mexico City and other large cities in the interior, where the great bulk of the country's population resides and where the concentration of industry is greatest.

Thus, because Mexico may be unable to afford control measures along the border, strengthened I/M programs for motor vehicles, or retrofits for existing, highly polluting sources, control **Scenario A** becomes a distinct possibility. With an increase of 63 to 305 percent in maquiladora activity in ten years, no-NAFTA industrial emissions under Scenario A may be expected to increase by 60 to 320 percent. With a concomitant estimated 40 to 180 percent increase in population in ten years, vehicular emissions may increase 40 to 200 percent, even taking into account a functional, centralized Mexican system of new automobile controls. Residential/commercial emissions may increase 40 to 200 percent, approximating population growth. The resulting composite Mexican border emission increases (40 to 225 percent) could be profound.

The second possible emissions outcome assumes that Mexico would somehow continue its aggressive border clean-up, as outlined in the Border Plan (**Scenario B**).

With a no-NAFTA, policy cooperation Scenario B future, assuming the same 63 to 305 percent Mexican border industrial growth, one could expect an increase in industrial air emissions in the 25 to 200 percent range. With an associated population increase of 40 to 180 percent but with an active vehicular I/M program and a federal new-car requirement for

EPA-like emissions standards, it is reasonable to expect a 40 percent decrease in vehicular emissions in ten years with the lower industrial growth rates; however, with higher industrial growth, vehicular emissions may increase by 90 percent. For residential and commercial sources, air emissions may increase approximately 15 to 140 percent, in spite of anticipated Mexican government efforts to subsidize cleaner-burning residential fuels. Thus, in ten years, a no-NAFTA, Scenario B future might result in changes in emissions ranging from a slight decrease (0 to 10 percent decrease) with the lower industrial growth projections to profound emissions increases (125 percent) with the higher industrial growth projections.

As Table 2 illustrates, reasonable estimates show that Mexican border emissions affecting the United States are likely to be less with NAFTA than without NAFTA.

b. Effects on the U.S. interior

In the absence of NAFTA, Mexico might implement automobile emissions standards comparable to the current U.S. standards; however, Mexico might not implement improvements to those standards as quickly. Moreover, the Mexican automobile fleet might not be renewed as quickly, I/M programs might not improve, and distribution of unleaded gasoline may remain limited. All of these influences would tend to result in a net increase in emissions from Mexican automobiles operated in the United States.

The situation is somewhat more complex regarding impacts of truck traffic from Mexico on air quality in the U.S. interior. Some consider it likely that the absence of a NAFTA would result in a continuation of the existing barrier to truck traffic beyond the border commercial zone. If this were the case, Mexican trucks would not affect interior U.S. air quality.

In the absence of NAFTA, the opportunity for increased cooperation through the North American Automotive Standards Council would be lost.

C. WATER USE AND QUALITY

Continued growth and development in the U.S.-Mexico border region -- which could occur with or without NAFTA -- will place increasing demands on already scarce water resources. There are three main areas of concern: (1) addressing limitations on surface water and groundwater supplies; (2) maintaining water quality through adequate wastewater treatment facilities; and (3) preserving ecosystems, wildlife habitats and coastal areas.

Implementation of the NAFTA will provide added impetus to cooperative projects already under way between the United States and Mexico, pursuant to the Integrated Border Environmental Plan, to promote water quality and preserve the border environment. Moreover, the Border Environment Cooperation Agreement will provide additional financing for infrastructure projects to treat wastewater and provide clean drinking water supplies.

Without NAFTA, the United States and Mexico may have difficulty sustaining their current level of cooperation in projects designed to maintain water quality. In addition, it may be difficult to obtain sufficient financing to design and construct needed wastewater treatment facilities in the border area. Moreover, in the absence of NAFTA, incentives for companies to locate facilities in the border region will continue, resulting in additional strains on already-scarce water resources.

1. Results of 1992 Recommendations

The 1992 Environmental Review included recommendations for areas of U.S.-Mexico cooperation on water quality issues, including exchanges of information on water regulations and technology, identification of water quality problems, and collection of data.

Since the 1992 report, EPA has continued its cooperative efforts with Mexico, in line with these recommendations. Pursuant to the Integrated Border Environmental Plan, the United States and Mexico are undertaking a number of initiatives to reduce water pollution and improve understanding of water resources in along the U.S.-Mexican border. Specific activities include:

- design and construction of new wastewater treatment facilities;
- initiation of pre-treatment programs for industrial wastewater;
- cooperative inventories of drinking water supplies; and
- surveys of toxic contamination of surface water supplies.

These activities are described in more detail in Section IV.A.

The Border Environment Cooperation Agreement (described in Section II.C) will provide new mechanisms to assist environmental infrastructure projects in the border region. Projects to provide wastewater treatment and clean drinking water supplies will be given priority consideration under this Agreement.

2. Background

Development and population growth in the U.S.-Mexico border area -- which could occur with or without NAFTA -- may have substantial effects on water quality and availability, especially given the fragile (or already degraded) nature of most of the ecosystems in the water-short desert areas. The degree to which the effects of development are mitigated will largely depend on the effectiveness of cooperative efforts between the United States and Mexico. These efforts must address three areas related to water quality and use:

- surface water and groundwater supplies;
- water quality and wastewater management; and
- watersheds, wildlife habitat, and coastal areas.

a. Surface water and groundwater supplies

Most of the larger communities along the U.S.-Mexico border obtain their drinking water from surface supplies, including the Rio Grande (Rio Bravo) and the Colorado River. Water quality is often poor, due to the presence of natural minerals, and generally declines as withdrawals increase.

Because aquifers provide base flows to local streams, aquifer depletion can have a significant adverse effect on surface and groundwater supplies. Agricultural or other industrial development in the border area which use groundwater supplies at rates greater than aquifer recharge will lead to depletion of groundwater reserves.

Water resources have been further strained by the growth of rural, unincorporated subdivisions in U.S. border counties (called "colonias"), which often have substandard or nonexistent water and sewer facilities. The General Accounting Office estimates that in Texas and New Mexico over 200,000 residents live in such colonias. In those without public water systems, residents typically use shallow wells that can be contaminated from private septic systems. While availability of drinking water is more an issue of water quantity than quality, lack of local financing capacity could result in poor drinking water quality.

The United States and Mexico are developing an inventory of the source, quality, and treatment processes for existing drinking water facilities of the sister cities. This inventory has been completed for cities on both sides of the border in the lower Rio Grande (Rio Bravo) valley. In addition, each government is in the process of determining its priority needs for water supply, treatment and distribution systems for existing and future development in the sister cities.

The U.S. and Mexican governments are also working to identify any areas where the drinking water sources common to both countries are contaminated, or where there is an identifiable threat of contamination. Efforts have begun to develop cooperative programs for solving problems that have already been identified.

In addition, the two governments are developing joint wellhead protection programs, designed to protect shared groundwater basins.

b. Water quality and wastewater management

In some areas of the border, the waters that cross the boundary or that are tributary to the international rivers present unacceptable sanitary conditions attributed to the disposal of wastewater in these watercourses. There is the related risk of pollution of transboundary groundwaters if proper management and treatment of surface wastewater is not carried out.

Wastewater disposal methods have an effect on health and the environment. The quality and quantity of effluents from wastewater treatment plants will affect the health of downstream surface water users, as well as the in-stream environment. In addition, groundwater may be adversely affected by wastewater land application or "lagooning" and by improperly managed solid waste disposal facilities. In colonias without sewers, residents typically use privies which do not meet public health standards and can degrade groundwater.

Municipal wastewater in the U.S. Sister Cities is treated by wastewater treatment facilities which are required to comply with discharge permit limitations. The United States and Mexico are currently working on plans to develop international wastewater treatment facilities for all the Sister City areas.

In several Sister City areas, efforts are under way to develop industrial wastewater pretreatment programs. Joint industrial inspections have been made for cross-border inspector training purposes. A course in how to establish an industrial wastewater pretreatment program is being prepared by EPA for use in Mexico.

Transboundary water quality monitoring is being increasingly well coordinated. For instance, a toxics monitoring study is being completed for the Rio Grande (Rio Bravo) and its tributaries on both sides of the border. The two governments are discussing extending these efforts to the Colorado River in the vicinity of the border.

Following are some examples of other recent cooperative efforts to address wastewater treatment needs in the Sister Cities:

- San Diego/Tijuana:

  The Tijuana wastewater collection system cannot convey and treat all of the wastewater generated in the city. As a consequence, some raw wastewater from Tijuana is being transported across the border, either by pipeline for treatment in San Diego on an interim basis or by flows down the Tijuana River to the Pacific Ocean.

  An international wastewater treatment plant is under design for the San Diego/Tijuana area, and preliminary construction has taken place.

- The New River:

  The New River, which originates south of Mexicali, flows north into California. The river carries both raw and partially treated domestic wastewater, industrial wastes and agricultural wastes. In California, additional agricultural runoff enters the river.

  Current plans for Mexicali/Imperial County wastewater include: achieving efficient operation of existing wastewater treatment lagoons; completing construction of new treatment facilities in southeast Mexicali to handle domestic and industrial wastewater from the area; elimination of all discharge of untreated domestic and industrial wastewater through expansion of the sewage collection system; and incorporation into the sanitary system of wastewater from new urban development related to the new Mexicali/Calexico port of entry. In 1992 the IBWC signed an agreement with Mexico (Minute 288) setting out a plan for projects to implement the above. Joint planning with Mexico is continuing. The Administration's funding request for U.S. participation is included in the FY 1994 budget.

- Nogales:

  Indications are that surface water in both the U.S. and Mexican Nogales has been contaminated intermittently with domestic wastewater, and there is possibly industrial and inorganic waste contamination.

  An expansion of the international wastewater treatment plant was completed in 1992. However, the expanded plant is expected to reach capacity in early 1994 and planning is in process for further expansion of the plant and provision of collection systems in Nogales, Sonora. The Administration has included funding for U.S. participation in this project in the FY 1994 budget. Discussions with Mexico continue on the development of a bi-national industrial pre-treatment program. This will be used as a model for other bi-national pre-treatment programs. Further, the United States and Mexico are planning the development of a bi-national groundwater monitoring and protection program.

- Nuevo Laredo:

  Construction of a wastewater treatment facility is under way.

In order to address the problem of inadequate wastewater treatment facilities in the U.S. colonias, infrastructure development programs are being established in Texas and New Mexico. These state programs are being initiated with a combination of federal and state funds. In addition, workshops have been held and a clearinghouse has been created for assisting people involved in managing colonia projects. Technology transfer seminars and demonstration project initiatives are also under way.

c. Watersheds, wildlife habitat and coastal areas

Estuaries and wetlands are a critical natural resource providing great economic, public health and ecological benefits. Over the long term, ecological degradation either directly or indirectly degrades human health and the economy.

Watersheds and wetlands are potentially threatened by growth. Since development generally follows water supply, the most attractive land is near streamflow, often in wetlands. Thus, growth could potentially have an adverse effect on local ecosystems and wildlife habitats.

Two areas of particular concern in the border area are the lower Rio Grande (Rio Bravo) valley and the San Diego Bay area:

- Rio Bravo Valley:

  The lower Rio Grande (Rio Bravo) valley is home to one of the most unusual ecosystems in the United States. Only about five percent of the original subtropical forest remains; the rest has been lost to agriculture, vacation homes and roads.

  The lower valley area is already experiencing problems with the discharge of wastewater into the river and into lagoons that eventually affect the river and the habitats in this area.

- San Diego Bay:

  Around the San Diego Bay, natural habitats have been greatly altered and reduced during the last 100 years. Nearly 90 percent of the salt marshes and 50 percent of the mudflats have been lost due to filling. Virtually all of the upland habitat has been converted to urban and industrial uses. The Bay supports an impressive number of species, including five endangered species, over 100 species of water fowl and shorebirds, and 90 species of fish and shellfish. The mudflats of the South Bay are a significant stopover for shorebirds of the Pacific Flyway.

  On the Mexican side of the border, an international wastewater treatment facility is being planned that will directly benefit the Tijuana River National Estuarine Reserve marshland. Since this facility will be located at the upstream edge of the Reserve, its outfall will bypass the Reserve.

3. Potential Impacts of the NAFTA

Implementation of the NAFTA should have several beneficial effects on water supplies in the border region. First, by reducing artificial incentives for companies to locate facilities near the U.S.-Mexico border, NAFTA may reduce the growth in demand for water in the border region.

Second, the Border Environment Cooperation Agreement associated with NAFTA will provide new financing for infrastructure projects to treat wastewater and provide clean drinking water supplies.

Finally, implementation of the NAFTA will lead to the establishment of the Commission for Environmental Cooperation, which will provide added impetus to U.S.-Mexican cooperation on managing water resources in the border area. Such cooperative efforts will become especially critical as reliance on international wastewater treatment plants is likely to increase. Since the character of transborder water problems is based on site-specific geographic and hydrological conditions, these problems can be most effectively addressed through cooperative efforts between U.S. and Mexican officials.

4. Likely Trends in the Absence of NAFTA

Without NAFTA, incentives for companies to locate near the border are likely to continue under the maquiladora program. It is possible that further development and population growth in the border area will be tempered by limitations on water availability. However, the potential for improved technology, water conservation and other factors makes the absolute limit on growth due to water limitations quantitatively difficult to predict.

To the extent that the rate of border growth could be expected to continue in the absence of NAFTA, the need for better management of water resources in the border area is likely to become more critical. For instance, increased industrial activity in areas dependent upon groundwater could potentially deplete local aquifers. In addition, as demand for water in the border area increases, water quality (and the need for additional wastewater treatment facilities) will become more important.

Without NAFTA, however, there might be some difficulty in sustaining the current level of cooperation between the United States and Mexico in providing and regulating water quality-related facilities. Moreover, it may be more difficult to obtain sufficient funding to finance the water treatment facilities which are so urgently needed in the border region.

D. CONTROL OF TOXIC CHEMICALS

Implementation of the NAFTA is likely to result in increased trade in chemicals between the United States and Mexico. Concerns have been expressed that the NAFTA could make it more difficult for the United States to enforce its laws and regulations designed to control trade in toxic chemicals.

The NAFTA maintains the ability of the United States to control imports of toxic chemicals from Canada and Mexico, pursuant to the Toxic Substances Control Act. In addition, the NAFTA includes provisions which encourage toxic chemical data and studies generated in Mexico to meet the same standards and Good Laboratory Practices as data generated in the United States.

1. Results of 1992 Recommendations

One of the key goals set out for U.S. negotiators in the 1992 Environmental Review was to ensure that "...the United States will maintain the right to exclude any products that do not meet its health or safety requirements."

As described in Section II, this goal has been met in the final texts of the NAFTA and the Environmental Agreement. The NAFTA contains explicit provisions that protect the right of the United States to determine its own appropriate levels of environmental protection. Moreover, the Enviornmental Agreement explicitly states that each country is free to determine its own levels of environmental protection, and commits the United States, Mexico and Canada to work to improve their environmental laws.

2. Background

a. Statutory framework

All commercial chemical substances (with the exception of pesticides, nuclear material, food, food additives, pharmaceuticals, cosmetics and medical devices) produced in or imported into the United States are regulated under the Toxic Substances Control Act ("TSCA"). TSCA provides authorization to control the manufacture, processing, commercial distribution, use and disposal of chemicals. Examples of the kinds of chemicals controlled by TSCA include (among many others) adhesives, surfactants, coatings, dyes, polymers, and chemical intermediates.

TSCA allows EPA to require testing of existing substances, review new chemical substances prior to commercial manufacture or import, control substances (e.g., by limiting or banning manufacture or use), and gather information. TSCA requires that exporters provide notification prior to exporting such substances, and that importers certify that the substances meet the provisions of the Act.

The EPA has developed detailed regulations specifying what information must be reported, as well as who is responsible for notification, data development and submission. Testing guidelines have been developed and published. These guidelines are often incorporated as test standards in rule making to facilitate the generation of health and safety data. In addition, Good Laboratory Practice ("GLP") regulations have been promulgated to ensure that studies will be carried out in an acceptable manner. These regulations, standards and GLPs provide reasonable assurance that the data generated are adequate and reliable for purposes of evaluating the potential risks of chemical substances.

b. International cooperation in generation of test data

The Organization for Economic Cooperation and Development ("OECD") has a testing guidelines program to be used by member countries when developing health and safety data. Member countries have agreed to accept data for assessment purposes which have been developed by other countries, as long as the data are generated using the OECD-approved test guidelines and are in accordance with GLP standards developed by the OECD.

As a member of the OECD, the United States has actively participated in the OECD program to develop these guidelines. The United States accepts data from other countries developed using either the OECD or TSCA guidelines. In fact, many of the OECD guidelines are based on TSCA guidelines. Mexico, with the support of Canada and the United States, has applied for membership in the OECD.

3. Potential Impacts of the NAFTA

Chapter Nine of NAFTA, which deals with standards-related measures, requires that such measures not discriminate between imported and domestically-produced products. TSCA meets the requirements of Chapter Nine, in that the Act, as well as its associated regulatory requirements, applies equally to domestically-produced and imported chemical substances.

In addition, Article 913 of the NAFTA establishes a Committee on Standards-Related Measures to facilitate compatibility of standards, consult regularly on matters on common concern, and enhance cooperation in developing, applying and enforcing standards-related measures. This article sets forth a number of specific topics for consideration by the Committee, including promotion and implementation of good laboratory practices and guidelines for testing of chemicals.

E. HAZARDOUS WASTE

With or without NAFTA, economic growth in the United States and Mexico is likely to generate increasing amounts of hazardous waste.

Article 104 of the NAFTA assures that U.S. agreements with Canada and Mexico to manage cross-border shipments of hazardous waste will prevail over NAFTA's obligations if there is an inconsistency. Article 104 makes the same provision for the Basel Convention, which governs international shipments of hazardous waste.

In addition, Chapter 12 of the NAFTA ensures that U.S. environmental engineering and waste management firms and professionals will be able to offer their services in Mexico to construct safe disposal facilities for hazardous waste.

In the absence of NAFTA, the amount of hazardous waste from Mexico requiring management in the United States could increase, as a result of two factors. First, the continuation of the maquiladora program is likely to result in increasing industrial activity in the border area, with associated growth in generation of hazardous waste. In contrast, implementation of the NAFTA could significantly reduce the incentives to locate in this region. Second, without NAFTA and its associated Border Environmental Cooperation Agreement, Mexico may have fewer funds to invest in waste management infrastructure.

1. Results of 1992 Recommendations

The 1992 Environmental Review included several recommendations for U.S.-Mexico cooperation on waste management. These included efforts to: identify the universe of waste generated at border facilities; exchange information; and conduct joint conferences and training.

Since 1992, the United States and Mexico have worked to establish compatible hazardous waste tracking systems. When fully compatible, these systems will enable the two governments to determine the amount of hazardous waste crossing the border and verify compliance with U.S. and Mexican laws concerning such shipments.

In addition, EPA has provided training to SEDESOL personnel in hazardous waste management and inspection techniques. EPA and SEDESOL also have conducted joint cooperative training visits to border facilities. The two agencies host an annual conference for maquiladoras and have jointly produced a bilingual manual concerning requirements for transboundary movements of hazardous waste.

The 1992 Environmental Review also made the following recommendations to the U.S. NAFTA negotiators:

- to protect rights and obligations under international environmental treaties (notably the Basel Convention); and

- to ensure that U.S. environmental engineering and waste management firms and professionals have access to the Mexican market.

These recommendations were followed by the U.S. negotiators and were substantially achieved in the final text of the NAFTA:

- Article 104 of the NAFTA states that provisions of the Basel Convention (which governs international shipments of hazardous waste) will prevail over NAFTA's obligations, in the event of any inconsistency, and an annex to Article 104 extends the same protection to U.S. bilateral agreements on hazardous waste with Canada and Mexico.

- Chapter 12 of the NAFTA secures rights of entry to the Mexican market for U.S. services providers, including companies and individuals offering waste management services.

2. Background

a. Existing controls for hazardous waste management

The Resource Conservation and Recovery Act ("RCRA") governs the management of hazardous waste in the United States. Under RCRA and its implementing regulations, the U.S. EPA has established a generation-to-disposal regulatory structure for hazardous wastes generated in or imported into the United States. These regulations include notification requirements, manifesting of hazardous wastes transported on public roads, and stringent permitting requirements for treatment, storage, and disposal facilities. Imported wastes are subject to all of the requirements that apply to wastes generated and managed in the United States.

The federal hazardous waste program in the United States is run in partnership with state governments, and the federal law can be supplemented by more stringent state laws.

b. U.S. capacity for hazardous waste management

The U.S. and Mexican governments have been working to improve the tracking of hazardous waste which must be returned from maquiladoras in Mexico to the United States for management and disposal. To date, not all of the waste generated by the maquiladoras can be accounted for. As a result, the United States does not have complete data on the amount of hazardous waste being returned to the United States from Mexico.

The available data, however, indicate that these shipments will have little effect on U.S. capacity needs. For example, EPA data for Region 6 (which includes the states of Texas and New Mexico) indicate that imports of hazardous waste through Texas border stations grew from 656 tons in 1987 to over 4,085 tons in 1992. These shipments represented a very small percentage of the total amount of hazardous waste generated in Region 6 (which amounted to over 66 million tons in 1987). The available data for Region 9 (which includes the states of Arizona and California) suggest similar trends.

Thus, based on the available information, the amount of hazardous waste currently entering the United States from Mexico for treatment and disposal appears to be extremely small relative to the total amount of hazardous waste generated and managed in the United States. Thus, these shipments appear to have an insignificant impact on U.S. hazardous waste management capacity.

c. Transboundary movements of hazardous waste

Transboundary movement of hazardous waste between the United States and Mexico is subject to the terms of a bilateral agreement, Annex III to the La Paz Agreement, which reflects many of the provisions of the Basel Convention. Most hazardous wastes entering the United States from Mexico are generated by maquiladoras. Mexican law generally requires that hazardous wastes generated at the maquiladora facilities from U.S. raw materials be returned to the United States for management. Under Annex III, the United States accepts the return of maquiladora waste for disposal consistant with U.S. standards. Under a Mexican Presidential Decree, imports of hazardous waste into Mexico are limited to wastes to be recycled.

International Requirements for Waste Exports. The Basel Convention on the Transboundary Movement of Hazardous Wastes and Their Disposal, which was negotiated under the auspices of the United Nations Environment Programme, was completed in March 1989. Over fifty countries have signed the Convention, including the United States, Mexico, and Canada. The Convention entered into force on May 5, 1992. Mexico and Canada have both ratified the Convention; the United States will ratify the Convention after passage of the necessary implementing legislation.

Many of the elements of the Convention are reflected in Annex III of the U.S.-Mexico Agreement on Cooperation for the Protection and Improvement of the Border Area Environment. Annex III to this Agreement, which was signed in 1983, specifies conditions for movements of hazardous wastes between the two countries. The United States entered into a similar agreement with Canada in 1986.

3. Potential Impacts of the NAFTA

Implementation of the NAFTA is likely to increase economic growth rates in the United States and Mexico which, in turn, is likely to increase the generation of hazardous waste. Given the current lack of adequate hazardous waste facilities in Mexico, increased generation of such wastes at maquiladora factories in Mexico could lead to increased waste capacity demand in the United States. This demand could be further increased by strengthened enforcement efforts on the part of both countries with respect to the maquiladoras, which could lead to an increase in the proportion of hazardous wastes which these facilities ship to the United States for treatment.

At the same time, as tariff barriers are eliminated under the NAFTA, there will be less incentive for firms to invest in the border area under the maquiladora program. As a result, there may be a decrease in the percentage of hazardous waste generated in Mexico which would be required to be returned to the United States. Such a reduction could offset to some degree any general increase in waste shipments to the United States resulting from economic growth in Mexico. In addition, the cost of hazardous waste management (including disposal) in the United

States is quite high, and is likely to continue to create strong incentives for waste producers to minimize the amount of hazardous waste generated.

However, there is little reason to expect that any increase in shipments of hazardous waste from Mexico would create significant adverse effects in the United States, since there appears to be sufficient waste management capacity to absorb any such increase. Furthermore, any increase in hazardous waste management would not be absorbed solely by border area states. Although the majority of hazardous wastes returned to the United States from maquiladoras are destined for treatment and disposal facilities near the border, significant amounts of these wastes are managed in other regions of the country.

4. Likely Trends in the Absence of NAFTA

While the potential adverse effects of NAFTA on hazardous waste management do not appear to be significant, the impacts in the absence of NAFTA could well be substantial.

In the absence of NAFTA, the maquiladora program will continue to provide incentives for companies to locate facilities along the U.S.-Mexican border. These facilities will generate increasing levels of hazardous wastes, which will need to be transported back to the United States for management.

In addition, without the NAFTA, Mexico may be unable to obtain adequate funding to build an effective regulatory infrastructure and waste management capacity in Mexico. As a result, Mexico will continue to depend on U.S. facilities to properly manage its hazardous wastes.

Finally, without additional cooperative programmatic and enforcement efforts, there is a greater likelihood that hazardous wastes imported into the United States will evade protective U.S. hazardous waste laws, and thus increase risks from potential mismanagement of these wastes.

F. NON-HAZARDOUS WASTE

With or without the NAFTA, economic growth in the United States and Mexico is likely to generate increasing levels of non-hazardous wastes. The lack of infrastructure and proper waste management practices along the border is already causing serious environmental and health concerns. New waste generation will only add to these problems.

The Border Environment Cooperation Agreement will help finance infrastructure projects addressing solid waste disposal needs. Moreover, implementation of the NAFTA may reduce the growth of waste along the U.S.-Mexico border by decreasing incentives for industries and populations to locate along the border.

In the absence of NAFTA, the continuation of the maquiladora program is likely to result in increasing population and industrial growth rates along the U.S.-Mexico border. Further, without NAFTA it may be very difficult to secure adequate funding to properly manage the waste generated in this region.

1. Results of 1992 Recommendations

The 1992 Environmental Review included several recommendations to address non-hazardous waste concerns. These included a needs assessment, waste collection improvements, and development of new landfills.

Since that time, the United States has compiled extensive data on municipal solid waste disposal facilities and completed an inventory of U.S. sites. Mexican authorities are currently conducting an assessment of municipal solid waste disposal facilities along the Mexican border area. In Nogales, a site for a new sanitary landfill has been selected.

The Border Environmental Cooperation Agreement (discussed in Section II.C) provides mechanisms for the U.S. and Mexican governments to support a wide range of environmental projects in the border region. Solid waste management facilities will be among the projects that have an initial priority under the Agreement.

2. Background

Rapid population growth in the border area in recent years has created severe pressures on municipal solid waste disposal facilities. In addition, there has been significant growth in industrial activity in the border area, which has resulted in increased generation of non-hazardous industrial wastes. Improper disposal and burning of non-hazardous wastes have created environmental risks in the border area from air and water pollution.

a. Municipal solid waste

Waste is generated in the border area of Mexico at the rate of 1.5 pounds per person per day. While this rate is lower than the U.S. national average of 4 pounds per person per day, the Mexican population generates more than 5,200 tons of municipal waste per day in this region, amounting to 1.9 million tons per year.

Of the total municipal solid waste generated in the border area of Mexico, EPA estimates that only about 40 percent (766,500 tons per year) is actually collected. This means that over 1 million tons per year of municipal solid waste are disposed of improperly. Moreover, most of the waste that is collected (about 500,000 tons per year) is disposed of in unlined, open-air dumps.

Unlined, poorly operated open dumps on the Mexican side of the border are a major environmental and health concern. These dumps may contaminate groundwater, and thereby pose significant exposure risks to the human population, since groundwater is a major source of drinking water supply along the border. In addition, such dumps have the potential to contaminate surface water in the United States.

Open air dumps have also created problems from air pollution resulting from burning of the waste. In Nogales, for example, open dumps may burn for days at a time, creating smoke which travels across the border to the United States.

An additional problem caused by population growth on both sides of the border has been the creation of unincorporated rural communities (known as "colonias" on the U.S. side of the border), which are characterized by substandard or nonexistent housing, roads, drainage, water and sewer facilities, and waste disposal services. Over 200,000 residents in Texas and New Mexico live in such communities. In the absence of adequate landfills, many communities have no way to dispose of wastes properly; as a result, wastes are dumped illegally. Existing landfill capacity in those communities with landfills may also become problematic in the near future. Finally, some municipal solid waste landfills on the U.S. side of the border cannot meet new landfill requirements under RCRA, and are being closed.

b. Industrial non-hazardous waste

The maquiladora program has resulted in increased industrial activity in the border area. While this has likely resulted in a significant increase in the volume of non-hazardous industrial waste generation, there is no current information available on the amount of non-hazardous industrial wastes generated or on how the waste is being disposed.

3. Potential Impacts of the NAFTA

The rate of municipal solid waste generation is directly related to population and income levels. However, it is difficult to forecast precisely the impact of the NAFTA on the already-rapid growth rates in the border area.

By decreasing the incentives for industries and populations to locate along the U.S.-Mexico border, the NAFTA may reduce the growth of waste in this area. Nevertheless, with or without NAFTA, continuing economic growth is likely to lead to increasing rates of waste generation, which could further strain existing landfill capacity in the border area. If proper controls are not implemented, further growth in the border region could lead to additional illegal dumping or burning of industrial wastes, with resulting air and water pollution affecting both Mexican and U.S. populations.

The Border Environment Cooperation Agreement accompanying the NAFTA will provide funding targeted at improving infrastructure and waste management programs along the border. Thus, this Agreement will address not only any increases in waste generation along the border, but also the existing environmental problems in that area resulting from waste mismanagement.

Moreover, increased economic growth rates resulting from NAFTA may well create additional revenues which could be used by municipalities to address solid waste disposal problems.

4. Likely Trends in the Absence of NAFTA

As noted above, population and industrial growth rates along the border are already rapid and are likely to continue to increase with or without NAFTA. In the absence of NAFTA, however, the incentives for industrial location near the border under the maquiladora program will continue or increase, which will create even greater pressures on waste disposal infrastructure in the border area. Moreover, without NAFTA, it may be difficult to secure the funding necessary to address both existing and future infrastructure needs for waste disposal.

G. CHEMICAL EMERGENCIES

Increased industrial growth near the U.S.-Mexico border -- which could occur with or without NAFTA -- may increase the likelihood of a chemical emergency affecting the environment of both countries.

The United States and Mexico are engaged in cooperative efforts to improve emergency preparedness coordination between the two countries. The Environmental Agreement specifically includes "environmental emergency preparedness and response activities" as an item in the work program of the Council for Environmental Cooperation.

Without the NAFTA, it is possible that the U.S. and Mexican governments would not be able to maintain the same level of commitment to such cooperative efforts.

1. Results of 1992 Recommendations

The 1992 Environmental Review included several recommendations for U.S.-Mexico cooperation with respect to chemical emergencies (pp. 135-136). These included: developing a joint response capability for the border; facilitating the establishment of local preparedness entities; and providing improved communications and training.

Since 1992, the United States and Mexico have completed a three-year workplan establishing schedules and priorities for joint contingency planning and emergency response. Joint exercises have been held for several Sister City contingency plans. In addition, EPA has provided training in hazardous materials awareness and workshops for public officials in the border communities.

The 1992 Environmental Review also included a recommendation that the NAFTA negotiators work to ensure a continued high level of support by the United States and Mexico for an effective emergency planning and response program for the border. The Environmental Agreement provides a list of issues on which the Council for Environmental Cooperation may wish to develop recommendations. Included on this list is "environmental emergency preparedness and response activities" (Article 10.2.k).

2. Background

The U.S.-Mexico Inland Joint Response Team ("Inland JRT") was created by the 1983 Environmental Agreement between the United States and Mexico. The Inland JRT was given responsibility for coordinating efforts in preparedness, mitigation, response and prevention related to hazardous substance releases along the inland border area.

The Inland JRT holds meetings annually (and more frequently when warranted) to address planning and preparedness issues. In January 1988, representatives of both countries completed and signed the U.S.-Mexico Joint Contingency Plan. In March 1989, the Inland JRT sponsored its first conference in San Diego. A second conference, focusing more specifically on the development of emergency response plans and joint response mechanisms along the border, was held in June 1990.

In 1993, the United States provided grants to the states of California, Arizona, New Mexico, and Texas to assist those states in developing Sister City contingency plans between cities along the U.S.-Mexico border. In addition, the U.S. government has given a grant to the International City/County Management Association ("ICMA"), in order to facilitate Sister City partnerships between U.S. and Mexico border communities through site-specific workshops and exchanging of technical assistance capabilities among Sister Cities.

Of the fourteen pairs of U.S. and Mexican Sister Cities (which are located across the border from each other) local joint hazardous substance emergency plans have been completed in Matamoros-Brownsville and Mexicali-Calexico. Plans are near completion for Nuevo Laredo-Laredo, Nogales-Nogales, and Tijuana-San Diego.

3. Potential Impact of the NAFTA

Implementation of the NAFTA will ensure that both the U.S. and Mexican governments maintain a high level of commitment to continuing cooperative efforts under the auspices of the Inland JRT. These efforts will help ensure that human health and the environment are protected from potentially devastating effects of chemical accidents, and that actions are taken to prevent such accidents.

In addition, as discussed in Section V.B.3 of this Report, implementation of the NAFTA should reduce the risk of chemical accidents by eliminating the need for transfer of hazardous materials from one truck to another at the U.S.-Mexico border.

4. Likely Trends in the Absence of NAFTA

Even without NAFTA, the industrial areas in and around the U.S.-Mexico border will continue to grow. Therefore, the United States and Mexico will continue to face the possibility of a chemical emergency which may have transboundary effects.

Although the Inland JRT and Sister City contingency planning efforts do not require large amounts of resources, the U.S. and Mexican governments must remain committed to the Inland JRT to ensure that its efforts are effective and to encourage the participation of private, as well as public, parties. However, in the absence of NAFTA, it is possible that the U.S. and Mexican governments would not be able to maintain the same level of organizational commitment and funding to cooperative efforts designed to prevent or mitigate chemical accidents.

H. WILDLIFE AND ENDANGERED SPECIES

Current development and activities in the U.S.-Mexico border area are having an adverse impact on wildlife and endangered species in the border area.

Although NAFTA could cause short-term slight to moderate increases in these adverse effects as a result of increased trade and development between the two countries, in the long term, increased opportunities for cooperation between the United States and Mexico as the result of the NAFTA will help to address the stresses of development. Furthermore, maquiladora development will tend to be dispersed away from the border area to other parts of Mexico, thus reducing its impact on the border area. Finally, as described in previous sections, the Environmental Agreement provides a mechanism to address many of these problems, especially those related to enforcement. In particular, new environmental funding and increased personnel could result in improved environmental conditions and reduced environmental effects in the border regions of both countries.

If NAFTA is not implemented, incentives will continue under the maquiladoras to locate facilities in the border areas, thus exacerbating environmental pressures on the border, such as loss of habitat, adverse impacts to endangered and threatened species, and reductions in groundwater levels. Moreover, many of the increased opportunities for cooperation, training and enforcement of wildlife protection laws discussed below would be lost.

1. General Effects

The implementation of NAFTA will have both positive and negative effects on wildlife, endangered species, and their habitats. These impacts will vary across the 2,000-mile border. For example, increased cooperation, technical assistance, and funding to remedy transboundary water quality problems may cause positive improvements for fisheries downstream, but could produce local negative effects on threatened and endangered species habitat if water treatment plants are constructed in riparian floodplain habitat. Closer coordination, as a result of NAFTA, could also reduce or eliminate local adverse effects.

More importantly, the effects of NAFTA on wildlife, endangered species, and their habitats will vary significantly over time. Some environmental problems that preceded NAFTA will likely continue after NAFTA is implemented. These include degradation of water quality, loss of riparian habitat, stress on threatened and endangered species and their habitat, and illegal traffic of listed wildlife plant and animal products.

In the past, development was poorly coordinated between the two countries with limited or no mitigation of adverse effects. NAFTA provides an opportunity for increased cooperation and coordination which may in both the short- and the long-term act to mitigate adverse effects of development on the border area, and enhance conservation in the region. The following discussion explores the probable short- and long-term effects of the NAFTA on endangered and threatened species, wildlife trade, migratory birds, wildlife habitat, refuges and fisheries.

Table 3 is a summary of wildlife and endangered species effects compared to existing conditions.

2. Wildlife Habitat and Biodiversity

There are many areas on both sides of the international border where habitat (including wetlands) is under pressure as a result of increased industrialization, infrastructure development, and agricultural development. For instance, it is estimated that ninety-five percent of the Rio Grande region's native habitat on the U.S. side of the border has been lost to agricultural and other development, and wildlife in the area is considered to be severely threatened. In addition, habitat in San Diego County and in the Tijuana area has been shrinking, primarily due to population growth. Illegal hunting is also threatening wildlife, including several endangered and threatened species. Ninety percent of Arizona's lowland riparian habitat has disappeared, and several species are threatened and need habitat protection.

In Mexico, the Pinacate Reserve in northern Sonora, a 49,000-acre park that is the most arid site in North America, is threatened by woodcutting, volcanic cinder mining, and poaching of the Sonoran pronghorn antelope, of which there are less than 100 remaining.

Wildlife habitat for migratory birds and endangered and threatened species has been and is continuing to be adversely affected by the construction of international bridges over the Rio Grande River valley. Where current bridges exist, new access roads are sometimes being built without adequate environmental review. Customs and related law enforcement facilities are occasionally built within the floodway, exacerbating the problem of wildlife habitat loss. Other threats to wildlife include contamination of the Rio Grande oxbow lakes, known as resacas. These small bodies of water serve as habitat for many species of birds, as well as fish, and used to be recharged naturally by the flooding of the Rio Grande. Pesticides also pose a menace to wildlife in the Rio Grande valley, having affected birds that eat contaminated insects and members of several bat species.

The increase in trade and corresponding agricultural, industrial, and commercial development pressures as a result of NAFTA could result in increased use and pollution of already stressed water resources, increased demand for sport fishing, and other recreational use of water resources. Because of the fragile nature of desert aquatic habitats, water management is a key concern.

With the expected increases in development resulting from NAFTA, there could be a moderate short-term increase in adverse impacts of the sort described above. These impacts, which can occur on both sides of the border, would decrease in the long-term due to the joint cooperation and funding of actions as a result of NAFTA to resolve environmental problems associated with existing, as well as future development pressure. The net long-term effect on wildlife habitat and biodiversity could be slightly positive.

TABLE 3

EFFECTS ON WILDLIFE AND ENDANGERED SPECIES

Wildlife and Endangered Species	Long-Term Effects Without NAFTA (5-10 years)	Short-Term Effects of NAFTA (1 - 5 Years)	Long-Term Effects of NAFTA (5 - 10 Years)
Migratory Birds	–	-	+
Wildlife Habitats and Biodiversity	—	–	+
National Wildlife Refuges	–	-	+
Endangered and Threatened Species	—	–	+
Wildlife Trade	—	–	+
Fisheries	–	-	+

KEY to Level of Effects

+++ high positive effect
++ moderate positive effect
+ slight positive effect
o no effect
- slight adverse effect
– moderate adverse effect
— high adverse effect

As incentives under the maquiladora program are eliminated as a result of the NAFTA provisions for trade and investment liberalization, industrial and population growth should be dispersed away from the border area. Associated impacts on wildlife habitat (including wetlands) and biodiversity should stabilize, thus providing an opportunity to improve existing conditions along the border to the benefit of both countries. Long-term positive benefits should be more dramatic on the Mexican side of the border than in the United States.

While construction of new or expanded bridge crossings in the lower Rio Grande valley could cause an increase in wildlife habitat loss, these short-term negative impacts could be mitigated through additional trans-boundary cooperation and local public involvement to allow new bridges and access roads to be elevated to allow a continuous corridor under the bridges. Customs and related facilities could be constructed outside of the floodway.

NAFTA can encourage the expansion of cooperative training programs with Mexico. Increased training of Mexican officials in biodiversity conservation and sustainable development, protection of habitat for wildlife and wetlands conservation, could result in an increase in conservation efforts and local support in Mexico. This, in turn, could result in increased long-term protection and enhancement of the natural resources in many areas of Mexico.

3. National Wildlife Refuges

Presently, serious habitat and wildlife management issues confront the eight refuges located along the border. Present adverse impacts to these refuges include effects from adjacent or upstream urbanization, industry, agriculture, contaminants, illegal Mexican immigration, drug traffic, and the U.S. efforts to control immigration and drug problems.

NAFTA could moderately increase, directly and indirectly, problems of refuges in the short-term, but could produce slight beneficial environmental effects for refuge resources in the long-term due to a commitment by both countries to resolve problems cooperatively. Such commitments are described in Section IV.

4. Endangered and Threatened Species

There are a number of endangered and threatened species in the border area due to activities affecting habitat on both sides of the U.S.-Mexico border. Endangered animal species include the jaguarundi, the ocelot, and the pronghorn antelope, while threatened plant species include the palmetto, the baretta tree, Johnston's frakenia, and ashy dogwood. About three dozen species of fish may already be extinct in the lower areas of the Rio Grande due to saltwater intrusion and pollution.

Passage of NAFTA, with increased trade across the border, could cause slight to moderate increased stress on threatened and endangered species in that area due to increased conversion of natural habitats to agriculture, and development of industrial facilities, highways, international bridges, and municipalities.

NAFTA could also result in slight positive, long-term benefits to threatened and endangered species on both sides of the border, due to increased cooperation, joint efforts, environmental education, training, and public outreach. Many species could benefit from NAFTA, as a result of increased cooperation efforts, including the Sonoran pronghorn antelope and desert tortoise.

5. Migratory Birds

Many species of neotropical migratory birds depend on wintering habitat in countries south of the United States. Many of these species are declining, due to forest fragmentation, tropical deforestation, and general habitat loss. Species using grasslands and open areas, as well as those with transitory requirements for wooded cover, are being adversely affected by heavy uses of pesticides. Most of these impacts are already occurring in the two border countries.

In the short-term, human development pressures in northern Mexico, as a result of NAFTA, could cause a moderate increase in adverse effects to migratory birds due to loss and degradation of riparian, floodplain, and montane forest from clearing, associated with industrial, commercial, and agricultural developments.

In the long-term, the opportunities for increased cooperation with Mexico on threatened and endangered species, refuges, pollution abatement, and other actions and programs, could result in slight positive effects on migratory birds and their habitats due to mitigation of ongoing and short-term impacts. NAFTA will promote an increase in the level of monitoring, research, management, and training in Mexico associated with the conservation of migratory birds and their habitats. Activities identified in Mexico's RESERVA program could also benefit migratory birds.

6. Wildlife Trade

The border area has been and remains a growing region of legal and illegal commercial plant and wildlife traffic for which detection and apprehension of illicit commerce is increasingly difficult. The increased flow of goods between the United States and Mexico under NAFTA could require increased law enforcement capability and an expansion of the infrastructure needed for communications, transportation, and other related services to address an expected increase in legal and illegal wildlife importation and commerce of wildlife products.

Actions to address the problem of increased trade are underway. In 1991, Mexico became a member nation of CITES. Through CITES, both the United States and Mexico can work more closely to increase wildlife enforcement along the border. Through NAFTA and the Environmental Agreement, both countries are encouraged to work closer on law enforcement, training, and educational projects; and additional funding will be made available for these activities. The net long-term effect should be reduced illegal trade, and thus, slight to moderate increased protection of endangered and threatened species.

I. FISHERIES

Although some fisheries-related environmental problems (in particular, the tuna-dolphin and shrimp-turtle problems) have been a source of friction between the United States and Mexico, Mexico has made substantial progress in addressing these problems. Indeed, Mexico has reduced its dolphin mortality rates dramatically.

NAFTA provides opportunities for increased cooperation in fishery management and conservation, and safeguards enforcement of laws relating to use of these resources, for example through the CEC established under the Supplemental Agreement. This could result in improved management of each species and conservation of depleted stocks.

Without NAFTA, current bilateral consultations will continue, but without the improved atmosphere and high-level attention that the Environmental Agreement provides. NAFTA has provided the incentive for many significant improvements in Mexico's conservation policies regarding living marine resources. Without it, the United States will not have a commitment from Mexico to give effect to these policies.

1. General Issues

All three NAFTA countries manage marine fisheries within exclusive economic zones 200 miles from their coasts. In accordance with generally accepted international practice, the coastal states manage resources under their jurisdiction for their benefit, but have the responsibility to conserve those resources. NAFTA does not affect the legal status of coastal state jurisdiction, its exclusivity, or such states' stewardship responsibility. Since obligations derived from fishery management laws are excluded from the dispute resolution provisions of the Environmental Agreement, challenges to the fishery management regime of each country will not occur under the Agreement.

Some fishery resources span or migrate across borders. While cooperation in the management of fisheries has been steadily increasing, particularly with Canada, the commitment to cooperation and mutual economic benefit in NAFTA supports an acceleration of coordination and exchange of data and information useful to each country's fishery management system. The United States believes in managing fisheries throughout their range. The enhanced cooperative atmosphere under NAFTA will facilitate that approach and the quality of fishery management decision-making affecting stocks of mutual interest.

Commercial exploitation of fishery resources will continue to be by national fleets; that is, NAFTA will not lead to a large influx of commercial fishing vessels of other parties fishing in one anothers' zones. The main source of new fishing pressure may come from increased tourism and associated recreational fishing.

The effect of NAFTA in terms of increased trade in fishery products and, by extension, on fishing will be driven by the fact that U.S. tariffs on most fishery products are already very low. Mexico already has good market access to the United States. On the other hand, new opportunities will open up for U.S.-produced fish. The increase is not expected to be large, however. Aquaculture is expected to grow in Mexico, stimulated by an improved investment climate owing to NAFTA, and may offer an alternative to fishing for species overexploited in the wild.

Currently, the extensive arid interior regions of Mexico are receiving pressure from a growing number of water users. This pressure is exerted on streams as water is withdrawn, diverted, and impounded for a wide variety of uses; as well as on aquifers due to increased withdrawal. Many of these uses also return polluted waters to the streams and rivers. As a result, interior fish populations are very stressed and limited in many areas. In some areas, increased human populations are also increasing sportfishing take of shell fish and finfish resources.

A slight, short-term increase in adverse effects to fishery resources could be expected due to NAFTA-induced economic growth. However, through NAFTA, increased cooperation, including funding and personnel for joint efforts could occur, thus reducing ongoing adverse effects and the immediate short-term adverse effects due to NAFTA.

In the long-term, NAFTA can result in slight, positive increases and more stable fishery populations along the border area, as increased water quality, wastewater treatment, and pollution control efforts are implemented between the two countries. Habitat improvements and protection efforts would also benefit fishery resources. NAFTA will directly benefit fishery management efforts, including a possible increase in cooperative fishery enforcement operations with Mexico, improved collection of baseline data through joint efforts, greater protection of water flows, and possible coordination with Mexico in management of shared resources on both coasts.

2. Specific Issues

The two bilateral fishery issues that have raised the most environmental concern with Mexico are the tuna-dolphin issue and the shrimp-turtle issue.

a. Tuna-dolphin

Since massive kills of dolphins in the Eastern Tropical Pacific ("ETP") fishery by U.S. vessels prompted Congress in 1972 to enact the Marine Mammal Protection Act ("MMPA"), the U.S. fleet has largely left the area. The Act prohibits the importation of yellowfin tuna caught with commercial fishing technology which results in the incidental kill of, or serious injury to, ocean mammals in excess of U.S. standards. Pursuant to the Act, importaton of yellowfin tuna caught in the ETP with purse seines by Mexican vessels has been prohibited since 1990.

In 1991, a GATT Panel requested by Mexico found the U.S. Marine Mammal Protection Act's embargo provisions to be inconsistent with GATT obligations. However, Mexico has been working with the United States to resolve the dispute and has not asked for adoption of the Report. In the meantime, the United States has worked within the GATT with other Contracting

Parties to find ways to ensure greater compatibility between international trade rules and environmental policies and objectives.

Meanwhile, the performance of the Mexican tuna fleet in reducing dolphin mortalities has improved dramatically. Dolphin mortality in the Mexican fleet (the largest fleet in the ETP tuna fishery) has declined from 51,000 in 1989 to a projected 1993 total of less than 2,000 dolphins. Thus far in 1993, the performance of the Mexican fleet in reducing the number of dolphins killed has been as good as, or better than, any fleet operating in the eastern Pacific, including the U.S. fleet whose performance had long been considered unmatchable. Since 1989, Mexico has reduced the number of dolphins killed in each tuna set from 8.5 in 1989 to two in 1992. Mexico has achieved these results through a combined effort by the government and industry including:

-- requiring the use of specific gear and techniques to reduce dolphin mortality to the lowest possible level;

-- mandating extensive training for captains and crews;

-- establishing strong penalties for tuna vessel captains and/or owners who violate regulations; and

-- requiring that every fishing vessel have an observer on board during every fishing trip to monitor compliance with required measures.

Both the United States and Mexico are active participants in the international program established by an Intergovernmental Agreement in 1992 and implemented under the auspices of the Inter-American Tropical Tuna Commission ("IATTC").

The Agreement has three principal components:

-- it establishes a per-vessel limit on mortalities that is reduced each year;

-- it requires an observer on every tuna purse seine vessel over a certain size operating in the fishery; and

- it establishes a research program to develop new fishing gear and techniques to reduce, and if possible eliminate, dolphin mortality in the fishery.

In addition, the Agreement establishes an International Review Panel (comprised of Government, industry, and NGO representatives) to monitor compliance and to recommend additional measures to ensure compliance with the Agreement.

The United States is committed to continued reduction of dolphin mortality with the goal of eliminating it through the most effective means available. The International Dolphin Conservation Act of 1992 authorized the implementation of a moratorium on encirclement of dolphins, subject to agreement by a major purse seine tuna fishing nation, but to date no nation has committed to such an approach. In June 1994, the United States will prohibit the importation and marketing of any tuna that is not shown to be dolphin-safe.

b. Sea turtles

Mexico's progress with respect to conservation of sea turtles has also been noteworthy, although there still are occasional verified reports of violations of conservation programs. Mexico has since 1991 implemented and enforced enhanced provisions to protect sea turtles caught incidentally in its shrimp-trawl fishery. More recently, Mexico in April of 1993 mandated the use of turtle excluder devices ("TEDs") on all commercial shrimp trawl vessels operating in the Gulf of Mexico and the Caribbean to eliminate the incidental capture and drowning of sea turtles in shrimp trawl nets. The Mexican shrimp fleet has received extensive training in the use of TEDs from U.S. gear technicians. That training will continue this fall for Mexican shrimp fishermen on the Pacific coast. With this latest action, Mexico will soon use TEDs on 100 percent of its vessels in the Caribbean and Gulf of Mexico before May 1994, when U.S. law would limit shrimp imports from nations not using TEDs. Recent reports indicate that mandatory TEDs training and stringent SEPESCA inspection will result in full compliance by the shrimp fleet with Mexico's TEDs program.

In addition, Mexico has for many years implemented programs that protect sea turtle nests on its beaches. An April 1990 Presidential decree set aside 17 nesting beaches as reserve and shelter zones for marine turtles. With the establishment of a National Marine Turtle Program, Mexico selected protection camps (beach sites) in both the Gulf and Pacific coasts where turtle eggs were studied, and protected through removal and relocation in special secure areas. In addition, this program has sought to promote turtle conservation through the protection of nests, fostering of artificial incubation of eggs where necessary, release of turtle hatchlings, and tagging of specimens. In 1990, Mexico also banned the previously legal harvest of olive ridley sea turtles in the state of Oaxaca. Mexico has also joined CITES and banned exports of sea turtle shell and skins to Japan and elsewhere. Mexico and the United States have cooperated for years on a program to save the highly endangered Kemp's ridley sea turtle whose only known nesting beach is at Rancho Nuevo, Mexico and have agreed to work together on a regional convention for sea turtle conservation.

3. Cooperative Efforts

One objective of the Environmental Agreement is to promote sustainable development based on cooperation and mutually supporting environmental and economic policies. Under this agreement, the United States will be able to consult with Canada and Mexico in the Commission for Environmental Cooperation ("CEC") on issues that the United States could previously discuss only on a more ad hoc basis for lack of a proper forum.

Fisheries enforcement cooperation could result in an improved effort to preserve endangered fish and cetacean species. An example of recent, improved enforcement cooperation between Mexico and the United States involves an agreement to protect the vaquita and the totoaba. Both vaquita and totoaba are endangered species that share habitat in the Sea of Cortez. Mexico prohibits by law the harvest of totoaba. The two countries have recently drafted an enforcement agreement to improve enforcement of the illegal harvesting on totoaba in Mexico. This is a positive cooperation effort made in anticipation of the NAFTA.

The CEC will also act as a forum for consultations on issues relating to the environmental implications of a good during its life cycle. The incidental death of vaquita during the harvest of totoaba, the incidental death of dolphin during the harvest of tuna and the incidental death of sea turtles during the harvest of shrimp are all examples of processes or production method ("PPM") problems that could in the future be addressed in the CEC. Consultations among the three countries on PPM issues will progress from the cooperation and exchange of science and technology that can occur during consultations to resolve that specific environmental concern.

In October 1993, the Agency for International Development ("AID"), with the National Marine Fisheries Service, agreed to provide the Government of Mexico and the Mexican Fisheries Ministry ("SEPESCA") with technical assistance and training for the protection of marine resources of binational importance, principally marine turtles and mammals taken in the shrimp and tuna fisheries.

Under the AID Agreement, NMFS will provide training, assistance and other technical support to the Government of Mexico and SEPESCA to strengthen their efforts in the use of TEDS, dolphin-safe research, biosphere reserve management and vaquita surveys. NMFS specialists will prepare final reports, including recommendations on these subjects. NMFS will also arrange to conduct a program evaluation during FY 1994 to determine training program impact, measuring progress toward specific activity targets and overall improvement in marine resource protection.

Marine pollution monitoring stations have been established throughout the United States, including seventeen along the U.S.-Mexico border, under the NOAA Status and Trends program. Sediments and/or biota have been sampled for several years to monitor pollution in the marine environment.

4. Marine Sanctuaries and Reserves

In order to protect valuable habitats for the marine mammal known as vaquita and the endangered finfish totoaba, President Salinas inaugurated a 1.7 million hectare biosphere reserve in the state of Sonora, on June 10, 1992. The two new biosphere reserves, located close by Puerto Penasco, are officially recognized as the Pinacate/Great Desert of Altair and the Upper Gulf of California/Colorado River Delta. The protected zone within the Gulf serves as home and breeding ground for the vaquita and the totoaba. Mexico also has a reserve in Baja, California for the grey whale to protect breeding habitat.

J. FORESTS, PARKS AND RANGELANDS

While deforestation in Mexico will likely continue, it is anticipated that NAFTA could slow the rate of deforestation somewhat, due to a general increase in wage rates and an increase in alternative sources of productive employment.

In the short term, resource threats to the National Park System may be created by an increase in cross border transportation, increased population settlement with some additional industrial site development at or near the border, consequent decreased water quality, and increased destruction of cultural resources. However, increased opportunities for joint cooperation to address these problems may alleviate some of their impacts. NAFTA may increase demand for services from public lands, but otherwise its impact on them will be negligible.

If NAFTA is not implemented it is unclear that alternate mechanisms to reduce deforestation exist. Pressure in the border areas would continue, particularly those driven by the concentration of maquiladoras. Without NAFTA, public use and demands on public lands will continue, but likely at lower levels than with the implementation of NAFTA. However, without NAFTA, the level of adverse impacts to park resources would be slightly greater than long-term conditions with NAFTA due to less joint coordination and cooperation between U.S. and Mexican environmental agencies to reduce environmental problems.

1. Forest Resources

Mexican forest resources are an important national and global resource. Many animal and plant species are endemic: 53 percent of the reptile species and 14 percent of the plant species in Mexico's forests exist nowhere else in the world. Mexico ranks seventh in tropical forest area. However, the U.N. Food and Agriculture Organization ("FAO") estimates Mexico is losing 1.3 percent of its forests per year. According to the World Wildlife Fund, regional studies suggest that half the forest cover has been lost in the last ten years.

A key cause of deforestation in Mexico is that most forested areas are considered an "open-access" resource. Although most forest land is nominally owned by ejidos (landowner associations) or the state, effective control is often contested. Hence, forest resources are likely to be overexploited. Furthermore, even in cases where control is not an issue, there is little incentive for forest resource users to consider the offsite effects such as increased siltation downstream or the loss of biodiversity. The recent change in the ejido system increases the uncertainty about ownership of Mexican forest resources and, at least in the short-term, speeds up the rate of deforestation.

A common misconception is that deforestation is the result of logging for timber production and exports. In Mexico, two other economic activities are more important: the collection of fuel wood (a cheap source of energy), and the conversion of forest land to agriculture (cropland and pasture). Poverty and unemployment play an important role in forest land conversion. The only resource of the poor is their labor. If wage rates are low or employment is unavailable, the poor must find other sources of survival. Collecting fuel wood and growing crops on previously forested land are often their only source of productive employment.

NAFTA will cause Mexican wage rates to rise and demand for labor in sectors with comparative advantage to increase. These two factors will provide the poor with new, productive employment and will reduce the pressure to collect fuel wood and convert forest land to crop production.

2. Impacts on National Park Service Areas and Responsibilities

The National Park Service, in line with its general responsibilities under the Archeological Resources Protection Act, anticipates a heavier law enforcement burden in controlling and investigating illegal excavations of archeological sites and illegal cross-border trade in archeological objects. This issue is of growing concern to the Native American community in both nations.

There are also concerns over the anticipated increases in cross border transportation as a result of NAFTA. The lack of a major east-west highway in Northern Mexico indicates potential pressures on current and possibly new north-south (cross border) corridors. An example is Arizona State Route 85 through Organ Pipe Cactus National Monument. This road, which already has been determined to have a detrimental effect on conservation of the park's natural resources, is currently a low-volume recreational use road. If volume increases, and evolves into heavy commercial vehicle use, negative impacts on the park will intensify. The proposed extension of I-17 from Flagstaff to I-15 has the potential for impacts on Glen Canyon National Recreation Area, Pipe Springs National Monument, and Zion, Bryce Canyon, and Grand Canyon National Parks. The latter, as a designated World Heritage Site, calls into play both U.S. and Mexican treaty obligations toward its protection.

Groundwater depletion is another area of ongoing concern. Organ Pipe Cactus National Monument, for example, is already documenting a significant draw down of the groundwater table due to irrigation of agricultural areas and increased pumping of groundwater for urban areas in Mexico. Also of concern to the park, in addition to ground water depletion, is the effect on native plants and animals of herbicide and pesticide drift, and invasion of non-native flora and fauna.

Another concern relates to plans for industrial site developments at or near the border. Concerns have recently been expressed by the National Park Service over Mexican construction of a new power plant (Carbon II) near an existing one (Carbon I) in the State of Coahuila. Both facilities are coal fired and lack pollution control equipment for sulphur and nitrogen oxides. There is significant potential for impacts on air quality, through visibility impairment and possibly negative effects on vegetation, in Big Bend and Guadalupe Mountains National Parks. Both areas are designated Class I under the Clean Air Act. (Recent developments on this issue are discussed in more detail in Section VI.B.)

Increased industrial development and urbanization at the border may also have impacts on park resources through water quality effects. The effects of upstream impoundments, channelization, diversions, and irrigation is impacting the Rio Grande, as it flows through Big Bend National Park and becomes the Rio Grande Wild and Scenic River. The above indicators indicate the probable emergence of additional and perhaps intensified short-term resource threat issues to National Park System units from activities reasonably to be anticipated as a result of NAFTA. However, many of these adverse threats could occur without the implementation of NAFTA as Mexico increases trade with the United States and economic development in its northern frontier regions and people are drawn to the border for jobs.

NAFTA promotes opportunities to improve the mechanisms for discussion and resolution of these disputes. The creation of a new North American Commission on Environmental Cooperation would facilitate the necessary technical/legal dialogue with Mexico, provide for independent arbitration of technical issues, and hold out the ultimate option of invoking trade sanctions on violators. No similar effective method of dispute resolution and settlement exists currently. In the long-term, the objectives of NAFTA and the Environmental Agreement should favor resolution of many of these issues.

3. Impacts on U.S. Bureau of Land Management Lands

Only minor impacts are expected to public rangelands along the U.S. border with implementation of NAFTA. There may be a slight increase in Mexican livestock grazing in the United States. However, this occurs quite regularly at present. Also, new livestock border crossings may be developed, as several are currently proposed.

Public lands along the border are receiving ever-increasing amounts of recreational use. Generally, recreational users of public land are providing a much needed economic boost to the small communities on both sides of the border. This recreational demand is expected to increase.

The U.S. Bureau of Land Management identifies and makes available utility corridors and road rights-of-way on request for rural communities. It issues free use permits for sand and gravel operations that support state and county road construction and maintenance. It also provides land to local governments at a nominal cost for recreation and public purposes. Through this program, the states, counties, and cities can acquire land for such things as parks, schools, wildlife management areas, and landfills.

It is expected that the border areas of New Mexico, Arizona, and California will prosper under NAFTA. This will result in increased requests from local governments for services that could be provided from public lands.

K. HEALTH IMPLICATIONS

Assuming that the NAFTA is successful in increasing economic development in Mexico generally, and along the border specifically, increased governmental support to the community and health infrastructure is expected, accompanied by an improved overall health status.

In the absence of NAFTA, if Mexico does not place greater emphasis on environmental enforcement along the border, the environmental health issues could increase geometrically, particularly for victims of surface water and groundwater pollution. Health problems ranging from bacterial infections to toxic effects from hazardous waste could take a dramatic upswing. Health and environmental resources, currently inadequate on both sides of the border, could be expected to remain so.

1. Background

The U.S.-Mexico border area population is increasing faster than the overall population of the United States, experiencing a 25 to 30 percent increase during the 1980s, compared with a 10 percent increase for the United States as a whole. Factors contributing to this rapid increase are high birth rates, lower age specific death rates, and high in-migration.

The population of the border counties is likely to be young, unemployed, and with a high percentage of persons living below the poverty level. The population on the border is younger than respective state and national averages, which will tend to result in a higher birth rate and increase the need for maternal and child care services. An expanding population of persons below the poverty level increases requirements for public resources and services.

Mexican women regularly cross the border to receive obstetrical services in the United States. There is also evidence that a large number of U.S. citizens go to Mexico for prenatal care, medical services and minor surgery, since medical care costs there are currently less. Data show that birth outcomes for these women, as measured by birth weights, are actually better than state averages.

Migrant and seasonal farm workers are of special concern in the U.S. border counties. The mobility of this population exacerbates the difficulty of obtaining accurate data, developing adequate planning and budgetary approaches, and implementing appropriate disease prevention and control services.

A significant public health threat resulting from increased migration to border counties is the spread of infectious disease. The high-population industrialized areas of the border, characterized by severe poverty, poor housing, crowded living conditions, environmental contamination and an absence of clean water and sanitation systems, are conducive to a high incidence of infectious diseases, particularly hepatitis, tuberculosis, measles and diarrheal diseases.

Unincorporated settlements exist on both sides of the border, with over 500 colonias along the Texas side exhibiting similar crowding and sanitation problems. Rates of amebiasis, shigellosis and campylobacteriosis in border counties are higher than for the State of Texas as a whole. Each of these diseases is caused by poor hygiene, polluted water, and contaminated foods, as is Hepatitis A.

Except for the County of San Diego, border counties are below the national average in numbers of all types of physicians, including primary care physicians. All of the border counties are categorized as Health Professional Shortage Areas under the HHS-funded National Health Service Corps funded by the Department of Health and Human Services ("HHS"). There are about 26 health professionals assigned through this program working in locations throughout the border counties of the four states.

Occupancy rates for hospitals in the border counties tend to fall below the national average, but it is not possible to determine whether this reflects an excess capacity of hospital beds or a deficit of health care personnel, particularly nurses, needed to operate at higher levels.

Community/Migrant Health Centers are federally funded primary care centers located in areas where medical service is insufficient. In the border counties a total of 50 clinic sites provide primary medical and dental care. These centers are serving medically underserved areas where the only other alternative for health care is often a hospital emergency room or no care at all.

The Department of Health and Human Services, through the Public Health Service ("PHS"), and Mexico's Secretariat of Health ("SSA") have been cooperating informally on border issues, on a regular basis, since 1942. This cooperation has been aided by the Pan American Health Organization ("PAHO"), which has provided a key informal forum for federal, state and local health officials to both discuss binational health issues and cooperate in intervention activities. Over the years cooperative activities have ranged from rabies control to tuberculosis prevention and control. Both epidemiological and health statistics data are provided to the PAHO Field Office by both sides. These data are of particular importance because they serve as an early warning about disease outbreaks as well as providing information about health status and health needs. The PAHO Field Office is also participating, in close cooperation with U.S. and Mexican federal, state and local governments, in the development of health profiles of border Sister Cities, to assist in both measuring the efforts of binational cooperation and in identifying emerging areas of need.

In a problem-oriented approach, PHS and SSA are cooperating in a new strategy focused on the U.S.-Mexico border. The program is limited to border Sister Cities, with responsibility for developing cooperative activities in the hands of city/county health officers. It emphasizes developing partnerships among governmental, academic and public and private for-profit and non-profit institutions. PHS and SSA are currently cooperating in a number of areas to facilitate the surveillance and control of cholera, and working in ten Sister Cities to develop binational health projects in six priority health areas.

Recently, HHS/PHS and EPA are giving increased attention to border environmental public health cooperation, including state and local health and environmental officials and volunteer organizations.

HHS/PHS and EPA are developing a memorandum of agreement covering a number of activities which would be carried out in conjunction with state and local officials. Under consideration are efforts to:

- gather and analyze information and data to develop baseline information on the current situation, which will allow the measurement of change over time.
- strengthen the statistical and epidemiological systems necessary to provide these data;
- develop a sentinel system to detect emerging, potentially environmentally-related illness or disease;
- conduct assessment of the risk of hazardous waste sites to local communities; and
- design and implement a program of research which can test results of research already conducted in other settings for relevance to the border situation; research to test questions of disease/illness occurrence identified through the sentinel system or epidemiologic surveillance; or research to identify environmental conditions that could impact human health.

2. Potential Impacts of the NAFTA

As noted above, there remain some formidable difficulties in forecasting with any accuracy the health implications of NAFTA implementation. Such an assessment must be based on assumptions of economic development, population growth, migration, environmental contamination, health status and other variables previously discussed.

In looking at the health implications for border communities, differences appear in the long term. Some projections anticipate a slowing of population growth in the border area in the near term, but it is not realistic to expect that resources will be immediately available to provide the infrastructure necessary to provide complete services for the border populations.

Should the type of industry that comes to the border change, or become more varied under the NAFTA, a different demographic pattern for the employed population and the risk in terms of occupational and environmental health effects could result. This could also result in changes in both disease patterns and demands for services.

In the longer term, however, implementation of the NAFTA is expected to be generally a positive influence on the situation on the border. The mix of border and interior industry growth should tend to keep up with the need for new jobs for border populations and keep migration down from the interior, due to employment opportunities closer to other urban centers. Assuming that the NAFTA is successful in increasing economic development in Mexico generally, and along the border specifically, increased governmental support to the community and health infrastructure is expected, accompanied by an improved overall health status.

Again, these predictions are difficult because so many unknown variables exist; however, socio-economic levels within communities usually have a negative relationship to health status. Under any scenario, the occupational health infrastructure will be inadequate in the foreseeable future to deal with the challenges. Even though some maquiladoras currently have health care personnel on the premises, they are usually not trained in occupational health. The focus thus far has been on participating in the delivery of health care services to workers and their families in cooperation with the Mexican Social Security System ("IMSS"). This focus can be expected to continue until improvements in the infrastructure provide the resources for delivery of necessary services to the total population.

3. Likely Trends in the Absence of NAFTA

Estimating the implications for health in border communities if the NAFTA is not implemented is also difficult, but in general less optimistic. It can be presumed that additional development will continue to take place on the Mexican side of the border, with the concomitant increase in migration from the interior of Mexico to the border. Environmental pollution will continue to be concentrated in existing highly populated corridors and the existing stresses on the infrastructure and services on both sides of the border will continue. The continued high rate of population growth will increase both the number of unincorporated settlements and the crowding of existing housing. Despite the current level of effort by U.S. and Mexican authorities to provide more sanitation, potable water services and basic health care, a larger absolute number of people remain without adequate services.

Without NAFTA, the key infectious diseases would continue to be major problems at the same incidence levels, particularly Hepatitis A, tuberculosis and the diarrheal diseases. Deteriorating sanitation/water supplies could result in cholera outbreaks in border communities which, thus far, have not occurred. Measles would continue to decrease as SSA vigorously implements childhood immunization campaigns throughout Mexico. Accidents and chronic diseases would increase slowly with the population increase and the absence of resources for effective prevention campaigns. Requirements for maternal and child health services, particularly obstetrical services, will increase on both sides of the border.

In the absence of NAFTA, if Mexico does not place greater emphasis on environmental enforcement along the border, the environmental health issues could increase geometrically, particularly for victims of surface water and groundwater pollution. Health problems ranging from bacterial infections to toxic effects from hazardous waste could take a dramatic upswing. Health and environmental resources, currently inadequate on both sides of the border, could be expected to remain so.

APPENDIX 1

BIBLIOGRAPHY OF NAFTA-RELATED STUDIES

Collins, Todd, Richard Hojel, Daniel Jinich and John Lonnquist. "Mexico, Free Trade, and the Environment: Will Mexico Become A "Pollution Haven" for U.S. Firms? A Study of Eight Industries." Harvard Business School, Harvard Field Study. May 1992.

Grossman, Gene M., and Alan B. Krueger, "Environmental Impacts of a North American Free Trade Agreement," NBER Working Paper No. 3914 (Cambridge, MA: National Bureau of Economic Research, November 1991).

Kellogg, Robert L., Margaret Maizel, and Don Goss. "Agricultural Chemical Use and Ground Water Quality: Where are the Potential Problem Areas?" U.S. Department of Agriculture, Soil Conservation Service, Economic Research Service, Cooperative State Research Service, and National Center for Resource Innovations. December 1992.

Pettyjohn, Wayne A., Mark Savoca, and Dale Self. "Regional Assessment of Aquifer Vulnerability and Sensitivity in the Conterminous United States." Robert S. Kerr Environmental Research Laboratory, Office of Research and Development, U.S. Environmental Protection Agency, 1991.

Reed, D. (Ed) (1992). "Structural Adjustment and the Environment." London: Earthscan Publication Ltd.

U.S. Environmental Protection Agency. "Evaluation of Mexico's Environmental Laws and Regulations: Interim Report of EPA Findings." August 1991.

U.S. Environmental Protection Agency. "Summary: Environmental Plan for the Mexican-U.S. Border Area (First Stage 1992-1994)." February 1992.

U.S. Environmental Protection Agency and SEDUE. "Integrated Environmental Plan for the Mexican-U.S. Border Area (First Stage 1992-1994)." February 1992.

U.S. Trade Representative's Office, "Review of U.S.-Mexico Environmental Issues." February 1992.